Parallel Power Electronics Filters in Three-Phase Four-Wire Systems

Man-Chung Wong · Ning-Yi Dai
Chi-Seng Lam

Parallel Power Electronics Filters in Three-Phase Four-Wire Systems

Principle, Control and Design

 Springer

Man-Chung Wong
Department of Electrical and Computer
  Engineering, Faculty of Science
  and Technology
University of Macau
Macao
China

Chi-Seng Lam
State Key Laboratory of Analog
  and Mixed Signal VLSI
University of Macau
Macao
China

Ning-Yi Dai
Department of Electrical and Computer
  Engineering, Faculty of Science
  and Technology
University of Macau
Macao
China

ISBN 978-981-10-9377-7   ISBN 978-981-10-1530-4   (eBook)
DOI 10.1007/978-981-10-1530-4

Printed on acid-free paper

This Springer imprint is published by Springer Nature
The registered company is Springer Science+Business Media Singapore Pte Ltd.

To my parents: Kok Kay and Yuk Ying
To my beloved wife: Ebbie Wong
To my daughter: Oriana Wong
Without your support, dedication
and understanding, this book could
have not been published

From Man-Chung Wong

To my parents: Xu-Yun Dai
and Xin-Feng Song
To my beloved husband: Rong Wang
To my sons: Roger Wang
and David Wang

From Ning-Yi Dai

To my parents: Sio-Hong Lam
and In-Fong Lei
To my beloved wife: Weng-Tong Cheong
To my brothers: Chi-Kit Lam
and Chi-Man Lam

From Chi-Seng Lam

Preface

Power Electronics is one of modern and key technologies in Electrical and Electronics Engineering for green power, sustainable energy systems, and smart grids. Especially, the transformation of existing electric power systems into smart grids is currently a global trend. The gradual increase of distributed generators in smart grids indicates a wide and important role for power electronic converters in the electric power system, also with the increased use of power electronics devices (nonlinear loads) and motor loadings, low cost, low-loss and high-performance shunt current quality compensators are highly demanded by power customers to solve current quality problems caused by those loadings. In distributed systems, three-phase four-wire systems are mainly installed for reducing the terminal phase voltage fluctuation. As a result, understanding the principles and operations of power converters in three-phase four-wire systems is important and helpful for Electrical and Electronic Engineers.

In this book, parallel three-phase four-wire power electronic converters are mainly focused. The content starts by reviewing the power quality issues, power quality standards, and comparing different power filter topologies, etc., in Chap. 1. In Chap. 2, the influences of reactive, harmonics, and unbalanced current problems on the electric power system are included; the instantaneous power theory and different coordinate control and transformations are discussed; different power electronic converter topologies and their corresponding space vector allocations are compared, which provides systematic, comprehensive, and detailed coverage of operation principles of three-phase four-wire power electronic filters. Chapter 3 is dedicated for active power filters (APFs), in which their corresponding circuits, mathematical models, 2D and 3D space vectors, pulse width modulations, DC voltage control, generalization control in multi-level algorithm and parameter design are described. Undoubtedly, lower cost, lower loss, and better performances of power converters will be the development trends and goals in these coming decades. Among different current quality compensators, hybrid active power filter (HAPF) possesses high potential to get balance amount cost, loss, and performance. In Chap. 4, with the comprehensive consideration of the cost, loss, performance, and anti-resonance capability, the design and control of a HAPF topology are given as the prospective solution for the low-cost, low-loss, and high-performance current quality compensation. In addition, the system performance analysis method, design, control techniques, system implementation, and their corresponding hardware platforms presented in this book can be extendable to other compensators and power converters.

The book is intended for researchers, Ph.D. students, postgraduates, and electrical power engineers who are specializing in power quality compensation, in which they can gain the specific knowledge of the design and control of parallel three-phase four-wire power electronics filters. Moreover, it is intended for bachelor students and postgraduates who are specializing in electrical engineering, in which they can gain the basic knowledge of current quality problems and its different compensating methods, power electronics converters, pulse width modulation (PWM), etc.

Man-Chung Wong
Ning-Yi Dai
Chi-Seng Lam

Acknowledgments

The research works presented in this book were carried out in the Power Electronics Laboratory/Electric Power Engineering Laboratory, Department of Electrical and Computer Engineering, Faculty of Science and Technology and State Key Laboratory of Analog and Mixed Signal VLSI, University of Macau (UM), Macao, China. We would like to express our gratitude to the Macao Science and Technology Development Fund (FDCT) with project codes: FDCT 098/2005/A, FDCT 015/2008/A1, FDCT 023/2009/A1, FDCT 109/2013/A3, and Research Committee (RC) of University of Macau (UM) with project codes: RG067/01-02S/HYD/FST, MYRG137(Y1-L4)-FST12-WMC, MRG012/WMC/2015/FST, RG013/0304S, RG015/0304S and RG062/0304S for funding these research works.

We would like to express our hearty and profound gratitude to Prof. Han Ying-Duo from Tsinghua University, China for having opened our vision. We also appreciate him for this immensely inspirational guidance and precious advice. We would like to thank Prof. Rui Paulo Martins for his great support and advice as well.

Dr. Man Chung Wong would like to express his appreciation to Prof. Zhou Li Gao for his supervision of his B.Sc. final year project about power factor correction. Dr. Wong started his power compensation research from that moment. And, Dr. Wong would like to thank Prof. Zhang Lin-Zheng for his help and guidance during his master study.

Last, but certainly not least, we would like to send our heartfelt appreciation to our families, who endured our dedication to this book.

Contents

1 **Introduction to Power Quality, Standards and Parallel Power**
  Quality Compensators 1
  1.1 Introduction 1
  1.2 Power Quality Issues 2
  1.3 Power Quality Standards 4
  1.4 Detailed Description of Current Quality Issues and Standards 6
      1.4.1 Harmonics 6
      1.4.2 Current Quality Standards 8
  1.5 Power Quality Measured Results in Macau 9
      1.5.1 Low Power Factor—Reactive Current Problem 11
      1.5.2 Current Harmonics 12
      1.5.3 Excessive Neutral and Unbalance Currents 12
  1.6 Shunt Power Quality Compensators 13
  1.7 Summary 15
  References 16

2 **Basic Principles for Parallel Power Electronic Filters** 19
  2.1 Parallel Compensation for Reactive Power,
      Current Harmonics and Unbalance 19
      2.1.1 Influence of Reactive Power 20
      2.1.2 Influence of Harmonics 21
      2.1.3 Influence of Unbalance 23
      2.1.4 Basic Principle of Parallel Compensation 25
  2.2 Instantaneous Power Compensation 29
      2.2.1 Definitions in a-b-c Frame 30
      2.2.2 Definitions in α-β-0 Frame 31
      2.2.3 Mixed-Coordinate Instantaneous Compensation 33

 2.3 Three-Phase Converters and Their Discussions 39
   2.3.1 Realization of 3 Dimensional Coordinates
      for Three-Phase Systems 40
   2.3.2 Basic Three-Phase Converters, 3D Coordinates
      and Their Comparisons 42
   2.3.3 DC Voltage Unbalance, Variation, Switching
      Functions and 3D Coordinates 52
 2.4 Summary 56
 References 56

3 **Active Power Filters** 59
 3.1 Development of Active Power Filters. 59
 3.2 A Two-Level Four-Leg VSI as a Three-Phase Four-Wire
   Active Power Filter 64
   3.2.1 Modeling of Two-Level Four-Leg Active
      Power Filters 64
   3.2.2 Voltage Control Signals According to the Required
      Compensating Currents 66
   3.2.3 Space Vector Analysis of a Four-Leg VSI 67
   3.2.4 Hysteresis PWM 71
   3.2.5 Space Vector Modulation 74
 3.3 Two-Level and Three-Level Three-Leg Center-Split VSI
   as Three-Phase Four-Wire Active Power Filters. 83
   3.3.1 Modeling of Three-Leg Center-Split Active
      Power Filters 84
   3.3.2 Space Vector Analysis of a Three-Leg
      Center-Split VSI 86
   3.3.3 Three-Dimensional Sign-Cubical Hysteresis PWM. 92
   3.3.4 Three-Dimensional Cylindrical Coordinate PWM 97
   3.3.5 Three-Dimensional Space Vector Modulation 105
   3.3.6 DC Linked Voltage Variation Control 114
 3.4 Three-Phase Four-Wire Multi-Level VSIs 123
 3.5 Generalized PWM for Multi-Level Three-Leg Center-Split
   and Four-Leg VSIs 125
   3.5.1 Generalized 3D Direct PWM 125
   3.5.2 Generalized FPGA-Based 3D PW Modulator 129
   3.5.3 Experimental Verification of the 3D PW Modulator. 136
 3.6 Design and Implementation of Active Power Filters. 137
   3.6.1 Minimum DC-Link Voltage Study for APF Under
      Reactive Power and Current Harmonics Compensation . . . 137
   3.6.2 Design of Coupling Inductor 146

        3.6.3   Implementation of a Three-Phase Four-Wire APF
                Prototype 151
        3.6.4   Experimental Results............................ 158
    3.7 Summary 163
    References 164
4   Hybrid Active Power Filters.............................. 167
    4.1 Development of Hybrid Active Power Filters (HAPFs)......... 167
        4.1.1   HAPF Topology 1—Series APF and Shunt PPF 168
        4.1.2   HAPF Topology 2—Shunt APF and Shunt PPF 169
        4.1.3   HAPF Topology 3—APF in Series with Shunt PPF...... 170
    4.2 Comparison Among Three General HAPF Topologies 171
    4.3 Existing Problems and Operation Principles of Conventional
         LC-HAPF... 173
        4.3.1   Existing Problems of Conventional LC-HAPF 174
        4.3.2   LC-HAPF Reference Compensating Current
                Determination Based on Single-Phase Instantaneous
                P-Q Theory 175
        4.3.3   LC-HAPF Reactive and Harmonic Reference
                Compensating Current Determination and PWM
                Control Block Diagram 176
    4.4 Analysis of Three-Phase Four-Wire LC-HAPF Compensating
         Performances 177
        4.4.1   LC-HAPF Single-Phase Harmonic Circuit Model 178
        4.4.2   Simulation Investigation of LC-HAPF Steady-State
                Compensating Performances...................... 182
    4.5 Dynamic Reactive Power Compensation and DC-Link
         Voltage Control Consideration for LC-HAPF 191
        4.5.1   Modeling of the DC-Link Voltage in a LC-HAPF
                Single-Phase Equivalent Circuit 192
        4.5.2   Influence on DC-Link Voltage During LC-HAPF
                Performs Reactive Power Compensation 194
        4.5.3   LC-HAPF Operation by Conventional DC-Link
                Voltage Control Methods......................... 197
        4.5.4   Proposed DC-Link Voltage Control Method 200
        4.5.5   Simulation and Experimental Verifications 205
    4.6 Adaptive DC-Link Voltage Control Strategy for LC-HAPF
         and APF in Reactive Power Compensation.................. 214
        4.6.1   Single-Phase Fundamental Equivalent Circuit Model
                of LC-HAPF 214
        4.6.2   LC-HAPF Required Minimum DC-Link Voltage
                with Respect to Loading Reactive Power 217
        4.6.3   Adaptive DC-Link Voltage Controller
                for a Three-Phase Four-Wire LC-HAPF.............. 220

4.6.4   Simulation and Experimental Verifications
              of the Adaptive DC-Link Voltage Controller
              for the Three-Phase Four-Wire LC-HAPF   224
      4.6.5   Simulation and Experimental Verifications
              of the Adaptive DC-Link Voltage Controller
              for the Three-Phase Four-Wire APF   233
4.7   Minimum Inverter Capacity Design for Three-Phase
      Four-Wire LC-HAPF   242
      4.7.1   Mathematical Modeling of a Three-Phase Four-Wire
              Center-Split LC-HAPF in A-B-C Coordinate   242
      4.7.2   Minimum Inverter Capacity Analysis of a Three-Phase
              Four-Wire Center-Split LC-HAPF   243
      4.7.3   Simulation and Experimental Verifications
              for Minimum Inverter Capacity Analysis
              of the Three-Phase Four-Wire LC-HAPF   246
4.8   Design and Performance of a 220 V/10 kVA Three-Phase
      Four-Wire LC-HAPF Experimental Prototype   251
      4.8.1   System Configuration of Three-Phase Four-Wire
              LC-HAPF   251
      4.8.2   Balanced and Unbalanced Testing Loads   252
      4.8.3   Design of Coupling LC of LC-HAPF   253
      4.8.4   Design of Active Inverter Part of LC-HAPF   254
      4.8.5   Experimental Results for a 220 V/10 kVA Three-Phase
              Four-Wire LC-HAPF Experimental Prototype   260
4.9   Summary   272
References   273

About the Authors

Man-Chung Wong received his B.Sc. and M.Sc. degrees in Electrical and Electronics Engineering from University of Macau, Macao at 1993 and 1997, respectively and Ph.D. degree in Electrical Engineering from Tsinghua University in 2003. He was a visiting fellow in Cambridge University, UK in 2014. He is an associate professor at University of Macau, China. His research interests include power electronics converters, pulse width modulation, active power filters, hybrid active power filters, and hybrid power quality compensator for high-speed railway power supply system. Most of the research results are published in the top journals such as IEEE transactions and IET journals. Until 2015, he has published more than 100 papers in the area of power electronics and six patents were granted in China and USA. Recently, industrial platforms were developed and installed in practical systems for power compensation based on his research results.

Professor Wong received several awards based on his research results such as Macao Young Scientific Award from Macau International Research Institute in 2000, Young Scholar Award of University of Macau at Year 2001, Second Prize for Tsinghua University Excellent Ph.D. thesis Award in 2003 and Third Class Awards of Macao Science and Technology Awards—Technological Invention Award in 2012 and 2014 respectively. Some of his postgraduate students received merit paper awards in conferences and champions in student project competitions.

He was selected to work in several conference committees and General Chair of IEEE TENCON 2015 Macau. He is with IEEE Macau Section and IEEE Macau Power Joint Chapter for many years. In 2014–2015, he was IEEE Macau Section Chair. Recently, he is North Representative of IEEE Region 10 Power and Energy Society and IEEE Macau PES/PELS Joint Chapter Chair.

Ning-Yi Dai received her B.Sc. degree in Electrical Engineering from the Southeast University, Nanjing, China, in 2001, and M.Sc. and Ph.D. degrees in Electrical and Electronics Engineering from the Faculty of Science and Technology, University of Macau (UM), Macao, China, in 2004 and 2007, respectively. From October 2007 to August 2009, she was a Postdoctoral Fellow of the Faculty of Science and Technology, UM. She is currently Assistant Professor in the Department of Electrical and Computer Engineering, University of Macau.

Dr. Dai has published more than 60 technical journals (IEEE-TIE, IEEE-TPEL, IEEE-TIA, IET-PE, etc.) and conference papers. She was co-author of five Chinese patents. Her research interests include power electronics converters, pulse width modulation, power quality conditioning for high-speed railway power supply system, and renewable energy integration and microgrid.

Dr. Dai was the co-recipient of the Macao Science and Technology Invention Award (Third-Class) in 2012. She received the 4th Regional Inter-University Postgraduate Electrical and Electronic Engineering Conference (RIUPEEEC) Merit Paper Award in 2006 and HuaWei Scholarship in 2000. Dr. Dai is an IEEE Senior Member. In 2015 and 2016, she was the Chair of IEEE Macau Women in Engineering Affinity Group.

Chi-Seng Lam received B.Sc., M.Sc., and Ph.D. degrees in Electrical and Electronics Engineering from the University of Macau (UM), Macao, China, in 2003, 2006, and 2012, respectively. From 2006 to 2009, he was Electrical and Mechanical Engineer in the Campus Development and Engineering Section, UM. In 2009, he returned to the Power Electronics Laboratory of UM to work as a Technician and at the same time started to pursue his Ph.D. degree in part time, and completed his Ph.D. in less than 3 years. In 2013, he was a Postdoctoral Fellow in the Hong Kong Polytechnic University, Hong Kong, China. He is currently Assistant Professor in the State Key Laboratory of Analog and Mixed-Signal VLSI, UM, Macao, China. He has co-authored two books: *Design and Control of Hybrid Active Power Filters* (Springer, 2014) and *Parallel Power Electronics Filters in Three-phase Four-wire Systems—Principle, Control and Design* (Springer, in press), one US patent, two Chinese patents and more than 50 technical journals (IEEE-TIE, IEEE-TPEL, IEEE-TIA, IEEE-PWRD, IET-PE, etc.) and conference papers. His research interests include integrated power electronics controllers, power management integrated circuits, power quality compensators, smart grid technology, electric vehicle charger, and renewable energy.

Dr. Lam was the co-recipient of the Macao Science and Technology Invention Award (Third-Class) in 2014 and the recipient of the Macao Science and Technology R&D Award for Postgraduates (Ph.D. Level) in 2012. He also received Macao SAR Government Ph.D. Research Scholarship in 2009–2012, Macao Foundation Postgraduate Research Scholarship in 2003–2005, the 3rd Regional Inter-University Postgraduate Electrical and Electronic Engineering Conference (RIUPEEEC) Merit Paper Award in 2005, BNU Affinity Card Scholarship in 2001 as well as Award of Dean Honor List of UM, in 1999–2000, 2000–2001, 2001–2002, 2002–2003, respectively.

Dr. Lam is an IEEE Senior Member. In 2007 and 2008, he was the GOLD Officer and Student Branch Officer of IEEE Macau Section. He was the Secretary of IEEE Macau Section during 2014–2016 and is the Vice-Chair of IEEE Macau Section in 2016–2018, and he is currently also the Secretary of IEEE Macau PES/PELS Joint Chapter (2015 IEEE PES High Performing Chapter Award). He was the Local Arrangement Chair of IEEE Region 10 Conference—TENCON 2015 and also ACM/IEEE Asia South Pacific Design Automation Conference—ASP-DAC 2016. Since 2009, he has also been a certified Facility Management Professional (FMP) (with Highest Score Prize awarded by Labour Affairs Bureau, Macao SAR and Macau Institute of Management) of International Facility Management Association (IFMA).

Abbreviations

A	Amperes
A/D	Analog-to-digital
AC	Alternating current
ADC	Analog-to-digital converter
APF	Active power filter
ASD	Adjustable speed drive
CB	Capacitor bank
CT	Coupling transformer
DC	Direct current
DPF	Displacement power factor
DSP	Digital signal processor
DVR	Dynamic voltage restorer
FFT	Fast Fourier transform
HAPF	Hybrid active power filter
HPF	High-pass filter
HVDC	High-voltage dc transmission
IEC	International Electro Technical Commission
IEEE	Institute of Electrical and Electronics Engineering
IGBT	Insulated gate bipolar transistor
IPM	Intelligent power module
L	Inductor
LC	Inductor and capacitor
LPF	Low-pass filter
P	Proportional
P.M.	Phase margin
PCB	Printed circuit board
PF	Power factor
PI	Proportional and integral
PPF	Passive power filter
PSCAD/EMTDC	Power system computer-aided design/electromagnetic transient in dc system
PWM	Pulse width modulation

RMS	Root mean square
S/H	Sample and hold
STATCOM	Static synchronous compensator
SVC	Static VAR compensation
THD	Total harmonic distortion
UPQC	Unified power quality compensator
UPS	Uninterruptible power supplies
V	Volts
VSI	Voltage source inverter

Chapter 1
Introduction to Power Quality, Standards and Parallel Power Quality Compensators

Abstract In this chapter, the introduction to power quality, standards and parallel power quality compensators are discussed. In Sects. 1.1–1.3, the overview of the power quality issues and the related international standards are reviewed. Then the detailed description of current quality issues are discussed in Sect. 1.4. After that, the power quality measurement results in Macau are reported in Sect. 1.5, which reveals three main current quality problems: reactive power, current harmonics and neutral current existed in the distribution power network. In order to relax those problems, different shunt power quality compensators: Capacitor Bank (CB), Passive Power Filter (PPF), Active Power Filter (APF) and Hybrid Active Power Filter (HAPF) can be adopted and their functionalities, pros and cons are also introduced in Sect. 1.6.

1.1 Introduction

We live in a world in which science and technology improves human living conditions. However, at the same time the improvement happened with the price of environmental degradation. The increase in demands of electric power generation for consumption might have increased the speed of the means of environmental pollution. Ways to reduce the energy resources, to optimize the power network, to generate power locally to reduce transmission loss and to adapt advanced power electronics and information technologies for the power network to be smart/plug-and-play are being investigated. This leads to the development of smart grid. This development is being accomplished not only for economic reasons but also for environmental reasons, in terms of reducing the cost of electricity and the interference among different loadings as well as improving the reliability of power supply. With the development of the smart grid, more and more renewable energy systems are being installed into the power networks, which may degrade the power quality supplied to the electric users. From the end users, more sensitive electronic appliances are applied to the grid, which may inject harmonics and reactive power into the power system. From the point of view of energy saving, reducing harmonics and reactive power can reduce extra unnecessary energy loss for electric operations and power transmission. As a result, it is mandatory to maintain the quality of power. In this chapter, power quality issues, its related standards, power quality measured results and parallel power electronics compensators are discussed.

1.2 Power Quality Issues

Based on the development of science and technology, the reformation of industrial structure and the recent development of smart grid technology, a higher demand on improved power quality is required. However, with the proliferation and increased

Table 1.1 Power electronics and motor applications [1]

(a) Residential	(e) Transportation
Refrigeration and freezes	Traction control of electric vehicles
Space heating	Battery charges for electric vehicles
Air conditioning	Electric locomotives
Cooking	Street cars, trolley buses
Lighting	Subways
Electronics (personal computers, other entertainment equipment)	Automotive electronics including Engine controls
Elevators, escalators	(f) Utility systems
(b) Commercial	High-voltage dc transmission (HVDC)
Heating, ventilating, and air conditioning	
Central refrigeration	Static VAR compensation (SVC)
Lighting	Supplemental energy sources (wind, photovoltaic), fuel cells
Computers and office equipment	
Uninterruptible power supplies (UPS)	Energy storage systems
Elevators, escalators	Induced-draft fans
(c) Industrial	Boiler feedwater pumps
Compressors	(g) Custom power devices
Blowers and fans	Active power filter (APF)
Machine tools (robots)	Dynamic voltage restorer (DVR)
Arc furnaces, induction furnaces	Unified power quality compensator (UPQC)
Lighting	
Industrial lasers	Static synchronous compensators (STATCOM)
Induction heating	
Welding	Uninterruptible power supply (UPS)
Elevators, escalators	(h) Aerospace
(d) Telecommunications	Space shuttle power supply systems
Battery chargers	Satellite power systems
Power supplies (dc and UPS)	Aircraft power systems

use of power electronics devices (nonlinear loads) and motor loadings, such as converters, adjustable speed drives (ASDs), arc furnaces, bulk rectifiers, power supplies, computers, fluorescent lamps, elevators, escalators, large air conditioning systems, and compressors, etc. [1–8], it is becoming more and more difficult to achieve this goal. Table 1.1 lists power electronics and motor applications that cover a wide power range from a few tens of watts to several hundreds of megawatts in residential, commercial, industrial and aerospace systems [1, 9].

On one hand, the widespread applications of power electronic devices enable the control and tuning of all power circuit for better performance, cost-effectiveness and enhanced energy efficiency. On the other hand, these applications increase the distortion and disturbances on the current and voltage signals in the power network. This is because the power electronic devices draw harmonic currents from the power utility and the harmonic voltage will then be generated, as the harmonic currents cause nonlinear voltage drops across the power network impedance. The presence of current and voltage components at other frequencies, and the negative and zero sequences in three-phase systems are harmful to the equipment of the power supply utilities and those of the customers. Harmonic distortion causes various problems in power network and consumer products, such as equipment overheating, capacitor fuse blowing, transformer overheating, mal-operation of control devices, excessive neutral current, degrades the defection of accuracy in power meters, etc. [5].

On the other hand, the usage of induction motor loadings will cause a phase shift between the current and voltage in the power network, resulting in the lowering of power factor of the loading. Loadings with low power factors draw more reactive current than those with high power factors as the power loss increases. Moreover, the larger the reactive current, the thicker cables are required for power transmission, which will either increase the cost or lower the transmission capacity of the existing cables. Thus, electricity utilities usually charge the industrial and commercial customers a higher electricity cost with low power factor situation.

All of these current and voltage phase shift and distortion phenomena are responsible for the deterioration of power quality in the transmission and distribution power systems. There is a need from both utilities and customers for power quality improvement. Consequently, power quality has become an issue that is of increasing importance to electricity consumers at all levels of usage. The principle circumstances resulting in the increased awareness of power quality issues can be attributed to several factors, which include [10–12].

• Conventionally, the load characteristic is inductive due to electro-mechanical loads. Recently, load characteristics have changed from electro-mechanical into computer applications. Large computer systems are used in many businesses and commercial facilities. These new equipment are more sensitive to power quality variations than the equipment applied before.
• Modern production manufacturers demand robust and stable power supplies. Any power quality issues happened means large economic loss especially in semiconductor production.

- The increasing emphasis on overall power system efficiency has resulted in a continued growth in the application of devices such as high-efficiency, adjustable-speed motor drives and shunt capacitors for power factor correction to reduce losses, resulting in increasing harmonic levels on power systems.
- Deregulation of power industry gives customers the right to demand higher power quality.

1.3 Power Quality Standards

There are two common international standards relating to the voltage and current distortion as shown below.

IEEE 519:2014: IEEE Recommended Practices and Requirements for Harmonic Control in Electrical Power Systems [10].
IEEE 1159:1995: IEEE Recommended Practice for Monitoring Electric Power Quality [11].

According to the above standards, Tables 1.2 and 1.3 summarize the voltage and current harmonics distortion limits. Table 1.4 shows the categories and typical characteristics of power system electromagnetic phenomena defined in IEEE 1159:1995.

The performances of power electronics compensators, electrical appliances or power supplies should reach the requirements of the above standards in order to achieve an acceptable power quality in the distribution and transmission power networks.

Table 1.2 Voltage harmonic distortion limits defined in IEEE 519:2014

Bus voltage at PCC	Individual voltage distortion (%)	Total voltage harmonic distortion THD_v (%)
69 kV and below	3.0	5.0
69.001 kV through 161 kV	1.5	2.5
161.001 kV and above	1.0	1.5

Note High-Voltage systems can have up to 2.0 % THD_v where the cause is an HVDC terminal that will attenuate by the time it is tapped for a user

Table 1.3 Current harmonic distortion limits

Standards	Total current harmonic distortion THD_i (%)
IEEE 519:2014	20 % THD_i (for small customers) 5 % THD_i (for very large customers)

Table 1.4 Categories and typical characteristics of power system electromagnetic phenomena defined in IEEE 1159:1995

Categories	Typical spectral content	Typical duration	Typical voltage magnitude
1. Transients			
1.1 Impulsive			
1.1.1 Nanosecond	5 ns rise	<50 ns	
1.1.2 Microsecond	1 us rise	50 ns–1 ms	
1.1.3 Millisecond	0.1 ms rise	>1 ms	
1.2 Oscillatory			
1.2.1 Low frequency	<5 kHz	0.3–50 ms	0–4 p.u.
1.2.2 Medium frequency	5–500 kHz	20 us	0–8 p.u.
1.2.3 High frequency	0.5–5 MHz	5 us	0–4 p.u.
2. Short-duration variations			
2.1 Instantaneous			
2.1.1 Sag		0.5–30 cycles	0.1–0.9 p.u.
2.1.2 Swell		0.5–30 cycles	1.1–1.8 p.u.
2.2 Momentary			
2.2.1 Interruption		0.5 cycles–3 s	<0.1 p.u.
2.2.2 Sag		30 cycles–3 s	0.1–0.9 p.u.
2.2.3 Swell		30 cycles–3 s	1.1–1.4 p.u.
2.3 Temporary			
2.3.1 Interruption		3 s–1 min	<0.1 p.u.
2.3.2 Sag		3 s–1 min	0.1–0.9 p.u.
2.3.3 Swell		3 s–1 min	1.1–1.2 p.u.
3. Long-duration variations			
3.1 Interruption sustained		>1 min	0.0 p.u.
3.2 Undervoltages		>1 min	0.8–0.9 p.u.
3.3 Overvoltages		>1 min	1.1–1.2 p.u.
4. Voltage imbalance		Steady state	0.5–2 %
5. Waveform distortion			
5.1 d.c. offset		Steady state	0–0.1 %
5.2 Harmonics	0–100th H	Steady state	0–20 %
5.3 Interharmonics	0–6 kHz	Steady state	0–2 %
5.4 Notching		Steady state	
5.5 Noise	Broad-band	Steady state	0–1 %
6. Voltage fluctuation	<25 Hz	Intermittent	0.1–7 %
7. Power frequency variations		<10 s	

1.4   Detailed Description of Current Quality Issues and Standards

In this book, only shunt power quality compensators are focused for solving the current quality issues. As a result, current quality issues and their related standards are discussed in this section.

1.4.1   Harmonics

Harmonics are sinusoidal voltages or currents having frequencies that are integer multiples of the frequency at which the supply system is designed to operate (e.g. 50 or 60 Hz). Harmonic distortion originates in the nonlinear characteristic of devices and loads in the power system. These devices can usually be modeled as current sources that inject harmonic currents into the power system. Voltage distortion results as these currents cause nonlinear voltage drops across the system impedance. Harmonic distortion is a growing concern for many customers and for the overall power system due to increasing application of power electronics equipment. The harmonic sources can be grouped into three categories according to their origin, size and predictability.

(1) Small and predictable

The residential and commercial power system contains large numbers of single-phase converter-fed power supplies with capacitor output smoothing, such as TVs and PCs. Although their individual rating is not significant, there is little diversity in their operation and their combined effect produces considerable odd-harmonics distortion. And the gas discharge lamps add to that effect since they produce the same harmonic components.

(2) Large and random

The most common and damaging load of this type is the arc furnace. Arc furnaces produce random variations of harmonics and interharmonics content which is uneconomical to be eliminated by conventional filters.

(3) Large and predictable

Large power converters, such as those found in smelters and HVDC transmission, are large harmonic current source and considerable thought is given to their local elimination in their design. When the a.c. system is weak and the operation is not perfectly symmetrical, uncharacteristic harmonics appear. While the characteristic harmonics of the large power converter are reduced by filters, it is not economical to reduce the uncharacteristic harmonics in that way and, therefore, even a small injection of these harmonic currents can, via parallel resonant conditions, produce very large voltage distortion levels.

Table 1.5 Results of harmonic performance test [18]

Load	Mode	Fundamental current (A)	THD-F (%)	Dominating harmonics	
Computer with monitor	On	0.54	110	3rd	58 %
Laser printer	Print	0.34	113	3rd	55 %
	Idle	0.11	160	3rd	52 %
Fax machine	Send	0.16	120	3rd	87 %
	Print	3.74	6	3rd	5 %
	Idle	0.11	98	3rd	54 %
Photocopier	Copy	5.56	26	3rd	20 %
	Idle	0.35	106	5th	42 %
UPS #1	Server	40	35	3rd	25 %
UPS #2	PC	4.3	130	3rd	89 %
Magnetic ballast w/cap	On	0.21	30	3rd	18 %
Electronic ballast #1	On	0.19	34	3rd	26 %
Electronic ballast #2	On	0.23	10	3rd	9 %
Sodium lamp	On	0.24	64	7th	44 %
Compact fluorescent lamp	On	0.1	136	3rd	49 %
Fan coil	On	8.5	5	5th	4.8 %
Lift	Run	39	36	5th	28 %

The main detrimental effects of harmonics are [12, 13]

- maloperation of control devices, signaling systems and protective relays,
- extra losses in capacitors, transformers and rotating machines,
- additional noise from motor and other apparatus,
- telephone interference.

In the past, concentration of the harmonic problem was put on the large rating industrial loads. But now more concentrations are put on the modern buildings, because large numbers of loads in these buildings are "dirty" [14–18]. Table 1.5 shows the result of harmonic performance test of various commonly used equipments [18]. Running a large number of these loads simultaneously will cause severe current harmonic problems especially in three-phase four-wire distribution systems.

It can be seen from Table 1.5 that the most troublesome harmonic for loads in modern buildings is the third harmonic. This harmonic in three-phase four-wire distribution system does not cancel each other but returns in the neutral, which can result in a current flowing in the neutral to exceed the line current. Excessive neutral currents in a system contribute to the following problems:

- neutral to earth voltages that create common-mode noise problems
- circulating currents flowing in transformers
- unbalance phase voltages
- high voltage drop at loads
- failure of the neutral conductor
- overheating of the neutral line

1.4.2 Current Quality Standards

There are large numbers of standards related to the power quality issues. Since current quality issues will be the focus in this book, only current quality standards will be discussed. Different countries developed their own harmonic standards and set their own emission limits according to their individual conditions and requirements. Some of the standards are listed in the following:

IEC 61000-2-1:1990
IEEE 519: 2014
IEC 61000-4-7: 1993
China National Standard GB/T14595-93

The most common harmonic index is Total Harmonic Distortion (THD). THD is defined as the r.m.s. of the harmonics expressed as a percentage of the fundamental component. However, THD value may be misleading when the fundamental load current is low. A high THD value for input current may not be of significant concern if the load is light because the magnitude of the harmonic current is low, even though its relative distortion to the fundamental frequency is high. To avoid such ambiguity, IEEE Standard 519:2014 defined a factor called the Total Demand Distortion (TDD) factor. The term is similar to THD except that the distortion is expressed as a percentage of rated or maximum load current magnitude, rather than as a percentage of the fundamental current. The mathematical expressions of THD and TDD are given as the following:

$$\text{THD:} \frac{\sqrt{\left(\sum_{n=2}^{50} I_n^2\right)}}{I_1} \tag{1.1}$$

$$\text{TDD:} \frac{\sqrt{\left(\sum_{n=2}^{50} I_n^2\right)}}{I_{rated}} \tag{1.2}$$

For the IEEE 519:2014, the harmonic current limits are based on the size of the load with respect to the size of the power system to which the load is connected. The ratio ISC/IL is the ratio of the short-circuit current available at the point of common coupling (PCC), to the maximum fundamental load current. The current distortion limits vary according to the system voltage level. The research work of this book will mainly focus on the user-end and the system voltage is smaller than 1000 V, therefore, the current distortion limits as listed in Tables 1.3 and 1.6 can be considered. For simplicity, $\text{THD}_i < 15\%$ for IEEE▲ can be considered as the current distortion limit to justify the compensating performances of parallel power electronic compensators in both simulations and experiments.
    ▲ With reference to the IEEE standard 519-2014 [10], the acceptable Total Demand Distortion (TDD) ≤ 15 % with I_{SC}/I_L is in 100 < 1000 scale (a small

Table 1.6 Current distortion limits for general distribution systems (120 V through 69000 V) in IEEE 519:2014

Maximum harmonic current distortion in percent of I_L						
Individual harmonic order (odd harmonics)						
I_{SC}/I_L	<11	$11 \leq h < 17$	$17 \leq h < 23$	$23 \leq h < 35$	$35 \leq h$	TDD
<20[a]	4.0	2.0	1.5	0.6	0.3	5.0
20 < 50	7.0	3.5	2.5	1.0	0.5	8.0
50 < 100	10.0	4.5	4.0	1.5	0.7	12.0
100 < 1000	12.0	5.5	5.0	2.0	1.0	15.0
>1000	15.0	7.0	6.0	2.5	1.4	20.0

Even harmonics are limited to 25 % of the odd harmonic limits above.

Current distortions that result in a dc offset, e.g., half-wave converters, are not allowed.

[a]All power generation equipment is limited to these values of current distortion regardless of actual I_{SC}/I_L

where I_{SC} = maximum short-circuit current at PCC

I_L = maximum demand load current (fundamental frequency comment) at PCC

rating experimental prototype). The nominal rate current is assumed to be equal to the fundamental load current at the worst case analysis, which results in THD = TDD ≤ 15 %.

1.5   Power Quality Measured Results in Macau

Macau is one of the two special administrative regions (SARs) of the People's Republic of China. Macau lies on the western side of the Pearl River Delta across from Hong Kong to the east, bordered by Guangdong Province to the north and facing the South China Sea to the east and south. The territory's economy is heavily dependent on gambling and tourism, but also includes manufacturing. The power quality measured results in Macau can be a typical example for one to figure out a draft picture of the power quality issues in different regions.

The power consumption distribution in Macau in 2004 and 2006 are shown in Tables 1.7 and 1.8, which are provided by the CEM (The Electricity Company of Macau) Statistics 2004 and 2006 [19]. It can be seen from Tables 1.7 and 1.8 that there is over 60.0 % of the total power consumption is from commercial buildings, hotels and recreation, and public administrative buildings.

In addition, Fig. 1.1 shows the total electricity consumption (GWh) in Macau between 1996 and 2012 [19]. Since the trend of the total electricity consumption (GWh) in Macau increases every year, and there are also more and more new casinos and hotels being constructed in Macau, the total power consumption from commercial buildings, hotels and recreation, and public administrative buildings in Macau will still continue to occupy a large portion of the total electricity consumption in future.

Table 1.7 Electricity consumption distributions in Macau in year 2004

Year 2004	Number of customers		Electricity sales	
Type of customer	Number	(%)	GWh	(%)
Domestic	173,760	87.1	588.9	31.3
Commercial, hotels and recreation, wholesale and retail	21,119	10.6	972.6	51.6
Industrial	2,422	1.2	141.9	7.5
Public sector and street lighting	2,281	1.1	180.7	9.6

Table 1.8 Electricity consumption distributions in Macau in year 2006

Year 2006	Number of customers		Electricity sales	
Type of customer	Number	(%)	GWh	(%)
Domestic	178,924	86.6	660	27.9
Commercial, hotels and recreation, wholesale and retail	22,802	11.0	1338	56.5
Industrial	2,378	1.2	150	6.3
Public sector and street lighting	2,410	1.2	220	9.3

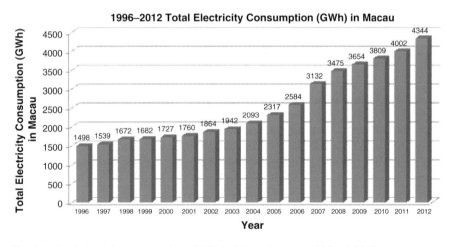

Fig. 1.1 Total electricity consumption (GWh) in Macau between 1996 and 2012

The commercials buildings, hotels and recreation, and public administrative buildings are generally equipped with large centralized air conditioning systems, lighting systems, elevators, escalators, computers and office equipment, etc. as shown in Fig. 1.2. Those facilities will cause power quality problems such as: reactive power, current harmonics and neutral current in the distribution power network. As a result, it is important to concentrate on the power quality problems of these facility types in Macau as well as in other parts of the world.

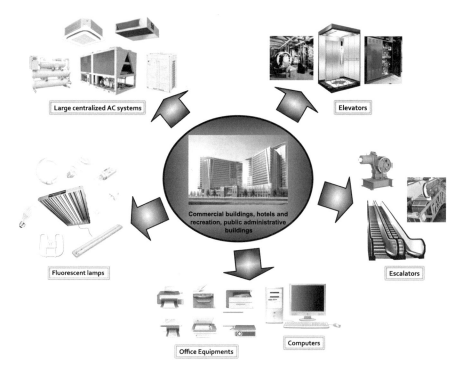

Fig. 1.2 General facilities installed in commercials buildings, hotels and recreation, and public administrative buildings

Since the Year 2003, the power quality monitoring works has been carried out by Power Electronics Laboratory (PELab), University of Macau in several Macau locations to evaluate the power quality situations. For the power quality measurement, a power quality analyzer ACE-4000, a survey-based power quality analyzer, is used. The power quality data is captured continuously for a complete working cycle (five business days and a weekend) at each measuring site. Based on the data that were measured and recorded for a commercial building, a hotel and recreation property, a middle school, an indoor sport center and three public administrative buildings, the most serious current quality problems are illustrated in the following sections such as low power factor, current harmonics, excessive neutral and unbalance currents. The details will be discussed in this section.

1.5.1 Low Power Factor—Reactive Current Problem

In Macau, there are three Tariff Groups according to the billing policy of CEM. Both customers of Group B and C need to pay for the reactive power when the power factor is lower than 0.857. Table 1.9 shows the power factor measurement

Table 1.9 Power factor measurement data for 7 buildings in Macau

Type of building	Power factor							
	Phase A		Phase B		Phase C		Three-phase	
	Max.	Min.	Max.	Min.	Max.	Min.	Max.	Min.
Commercial building	0.953	**0.808**	0.950	**0.825**	0.915	**0.758**	0.940	**0.801**
Hotel	0.997	0.910	0.996	0.911	0.993	0.887	0.994	0.905
Middle school	0.953	**0.761**	0.996	**0.750**	0.969	**0.627**	0.967	**0.771**
Indoor sport center	**0.686**	**0.634**	**0.809**	**0.748**	0.901	**0.854**	**0.771**	**0.723**
Public administrative building A	0.963	**0.792**	0.972	**0.791**	0.962	**0.807**	0.957	**0.797**
Public administrative building B	0.885	**0.735**	0.879	**0.727**	0.922	**0.758**	0.887	**0.742**
Public administrative building C	0.959	**0.724**	0.967	**0.724**	0.937	**0.714**	0.951	**0.720**

data for these 7 buildings in Macau [20], the **bold** data illustrates that the corresponding building is required to pay extra fees for the reactive power consumption.

1.5.2   Current Harmonics

From the current distortion limits in Table 1.3, nearly all the recorded THDi of phase currents for these 7 buildings in Macau exceed the limit. The THDi measurement data for these 7 buildings in Macau are summarized in Table 1.10 [20].

1.5.3   Excessive Neutral and Unbalance Currents

This severe problem has been overlooked for a long time. However, it is not uncommon for the neutral current to exceed the phase current, which gives rise to a

Table 1.10 Total current harmonic distortion (THD$_i$) measurement data for 7 buildings in Macau

Type of building	Total current harmonic distortion (THD$_i$ %)					
	Phase A		Phase B		Phase C	
	Max.	Min.	Max.	Min.	Max.	Min.
Commercial building	**37.04**	9.08	**39.96**	9.61	**45.41**	9.67
Hotel	**33.21**	6.04	**30.80**	7.14	**28.66**	6.47
Middle school	**46.83**	3.01	**64.13**	3.54	**63.73**	2.89
Indoor sport center	13.75	10.89	**17.89**	14.40	**24.47**	18.79
Public administrative building A	**28.09**	3.92	**50.75**	3.78	**47.65**	5.60
Public administrative building B	**19.27**	2.82	**20.06**	2.42	**27.12**	3.56
Public administrative building C	**25.61**	2.54	**28.84**	3.10	**27.01**	2.37

Fig. 1.3 Phase and neutral current RMS measurement data for one building

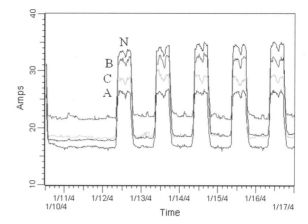

possibility of overloading the neutral conductor, thus also having a potential of causing fire. Moreover, it will increase transformer loss. Figure 1.3 shows the phase and neutral current root mean square (RMS) measurement data for one building [20], in which the neutral current is always larger than the phase current during office hours and the currents are unbalanced.

From Tables 1.7 and 1.8, there are over 60 % of the total power consumption from commercial buildings, hotels and recreation, and public administrative buildings. They are usually equipped with a large centralized air-conditioning system. If the loading is dominated by a centralized air-conditioning system, its reactive power consumption will be much higher than the harmonic power consumption [20]. As a result, the reactive power problem is usually more serious than the harmonic power problem if the commercial building or hotel and recreation or public administrative building is equipped with a large centralized air conditioning system. Therefore, the reactive power problem will be the first priority to be solved in Macau. In order to solve the current quality problems, shunt power quality compensator can be implemented between the load and the power supply sides. In the following section, the development and different shunt power quality compensators in distribution power system are introduced and discussed.

1.6 Shunt Power Quality Compensators

Shunt capacitor banks (CBs) as shown in Fig. 1.4 are used extensively in distribution power systems for power-factor correction and feeder voltage control. The principal advantages of CBs are their low cost and flexibility of installation and operation. However, the CB can easily be burnt if the current harmonics level is high. Moreover, the reactive power output is reduced at low load voltage. In order to solve the current harmonics problem, passive power filters (PPFs) as shown in Fig. 1.5 can be employed. Since the first installation of PPFs in the mid 1940s,

Fig. 1.4 Parallel capacitor
bank (CB)

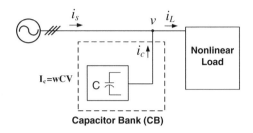

Fig. 1.5 Passive power filter
(PPF)

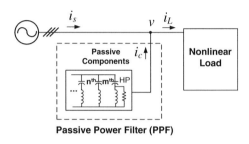

Fig. 1.6 Active power filter
(APF)

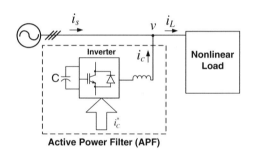

PPFs have been widely used to suppress current harmonics and compensate reactive power in distribution power systems [21] due to their low cost, simplicity and high efficiency characteristics. Unfortunately, there are many disadvantages such as low dynamic performance, resonance problems, filtering characteristic easily be affected by small variations of the system parameters, etc. [3–8, 22, 23]. Since the concept "Active ac Power Filter" was first developed by Gyugyi in 1976 [4, 21], the research studies of the active power filters (APFs) as shown in Fig. 1.6 for current quality compensation prosper. APFs can overcome the disadvantages inherent in PPFs, but their initial and operational costs are relatively high [3–7, 24, 25] due to the costs of semiconductor switching devices with its drivers and digital controller. In addition, the dc-link operation voltage of APF should be higher than the load voltage, thus the cost and switching loss of the switching devices would increase. As a result, their large-scale application in distribution power networks slows down.

Fig. 1.7 Hybrid active power
filter (HAPF)

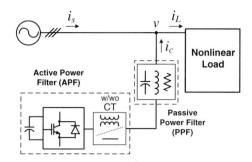

Table 1.11 Operational functions for shunt current quality compensators in distribution power system

		Shunt current quality compensators			
		Passive components		DFACTS	
		CB	PPF	APF	HAPF
Current quality	Current harmonics	–	**	**	* or **(D)
	Neutral current	–	–	*	*
	Reactive current	**	*	*	**(D) or *
		–	–	N	N

The symbol '*' means that the device has this function
The symbol '**' means that this is the main function for this device
The symbol '**(D)' means that this is the main function for this device if design accordingly
The symbol 'N' means that the device needs non-sinusoidal waveform
CB Capacitor Banks, *PPF* Passive Power Filters
APF Active filters, *HAPF* Hybrid Active Power Filters

Later on, different Hybrid Active Power Filter (HAPF) topologies composed of active and passive components in series and/or parallel have been proposed [3–8, 26–34] aiming to improve the compensation characteristics of PPFs and to reduce the voltage and/or current ratings (costs) of the APFs, thus leading to the relatively effectiveness in system cost and performances. Figure 1.7 shows a circuit config-uration of a typical type of HAPF. Table 1.11 summarizes the operational functions for these four current quality compensators [32–35] in distribution power network.

1.7   Summary

In this chapter, the introduction of power quality issues, standards, measured results in Macau and the power quality compensators are given. In the upcoming chapters, basic principles, operational algorithms and characteristics for parallel power electronics filters: active power filters (APFs) and hybrid active power filters (HAPFs) are discussed respectively.

References

1. B.K. Bose, Power electronics—a technology review. Proc. IEEE **80**(8), 1303–1334 (1992)
2. H. Akagi, New trends in active filters for power conditioning. IEEE Trans. Indus. Appl. **32**, 1312–1322 (1996)
3. H. Fujita, H. Akagi, A practical approach to harmonic compensation in power systems – series connection of passive and active filters. IEEE Trans. Indus. Appl. **27**, 1020–1025 (1991)
4. F.Z. Peng, H. Akagi, A. Nabae, A new approach to harmonic compensation in power systems —a combined system of shunt passive and series active filters. IEEE Trans. Ind. Appl. **26**, 983–990 (1990)
5. L. Chen, A.V. Jouanne, A comparison and assessment of hybrid filter topologies and control algorithms, in *Proceedings IEEE 32nd Annual Power Electronics Specialists Conference, PESC. 01*, vol. 2, 2001, pp. 565–570
6. F.Z. Peng, H. Akagi, A. Nabae, Compensation characteristics of the combined system of shunt passive and series active filters. IEEE Trans. Indus. Appl. **29**, 144–152 (1993)
7. P. Salmerón, S.P. Litrán, A control strategy for hybrid power filter to compensate four-wires three-phase systems. IEEE Trans. Power Electron. **25**, 1923–1931 (2010)
8. P. Salmeron, S.P. Litrán, Improvement of the electric power quality using series active and shunt passive filters. IEEE Trans. Power Del. **25**, 1058–1067 (2010)
9. N.Y. Dai, A generalized 3D pulse width modulator for multi-level voltage source inverters in three-phase four-wire power systems. Ph.D. thesis, University of Macau, Macau SAR, 2007
10. IEEE Recommended Practices and Requirements for Harmonic Control in Electrical Power Systems, 2014, IEEE Standard 519-2014
11. IEEE Recommended Practice on Monitoring Electric Power Quality, 1995, IEEE Standard 1159-1995
12. J. Arrillaga, N.R. Watson, S. Chen, *Power System Quality Assessment* (Wiley, New York, 2000)
13. E. Conroy, Power monitoring and harmonic problems in the modern building. Power Eng. J. **15**(2) (2001)
14. T.Q. Tran, L.E. Conrad, B.K. Stallman, Electric shock and elevated EMF levels due to triplen harmonics. IEEE Trans. Power Del **II**(2) (1996)
15. C. Boonseng, W. Koykul, S. Payakkaruang, M. Chikine, S. Kaewrul, The future growth trend of neutral currents in three-phase computer power systems caused by voltage sags, in *Power Engineering Society Winter Meeting, IEEE*, vol. 2, 23–27 Jan. 2000, pp. 1416–1421
16. H.O. Aintablian, H.W. Hill, Harmonic currents generated by personal computers and their effects on the distribution system neutral current, in *Industry Applications Society Annual Meeting*, vol. 2, 1993, pp. 1483–1489
17. T. Fukami, T. Onchi, N. Naoe, R. Hanaoka, Compensation for neutral current harmonics in a three-phase four-wire system by a synchronous machine, in *IEEE International Electric Machines and Drives Conference, IEMDC 2001*, 2001, pp. 466–470
18. T.K. Chiang, C.K. Law, V. To, H.F. Kwan, Power quality case studies: CLP power's experience, in *Proceedings of Symposium on Power Quality and You: Managing Pollution in Electric Supply Systems*, 2002, May
19. Annual Reports, The Electricity Company of Macau, 2004 and 2006. www.cem-macau.com
20. S.U. Tai, Power quality study in Macau and virtual power analyzer. Master's thesis, University of Macau, Macau, 2012
21. S.T. Senini, P.J. Wolfs, Systematic identification and review of hybrid active filter topologies, in *Proceedings IEEE 33rd Annual Power Electronics Specialists Conference, PESC. 02*, vol. 1, 2002, pp. 394–399
22. J.-H. Sung, S. Park, K. Nam, New hybrid parallel active filter configuration minimising active filter size, in *IEE Proceedings Electronics Power Application*, vol. 147, no. 2, pp. 93–98, March 2000

23. D. Rivas, L. Moran, J.W. Dixon, J. Espinoza, Improving passive filter compensation performance with active techniques. IEEE Trans. Indus. Electron. **50**, 161–170 (2003)
24. S. Khositkasame, S. Sangwongwanich, Design of harmonic current detector and stability analysis of a hybrid parallel active filter, in *Proceedings of Power Conversion Conference*, vol. 1, 1997, pp. 181–186
25. Z. Chen, F. Blaabjerg, J.K. Pedersen, Harmonic resonance damping with a hybrid compensation system in power systems with dispersed generation, in *IEEE 35th Annual Power Electronics Specialists Conference, PESC 04*, 2004, vol. 4, pp. 3070–3076
26. S. Bhattacharya, D. Divan, P. Cheng, Hybrid solutions for improving passive filter performance in high power applications. IEEE Trans. Indus. Appl. **33**, 732–747 (1997)
27. H. Fujita, T. Yamasaki, H. Akagi, A hybrid active filter for damping of harmonic resonance in industrial power systems. IEEE Trans. Power Electron. **15**, 215–222 (2000)
28. H. Akagi, S. Srianthumrong, Y. Tamai, Comparisons in circuit configuration and filtering performance between hybrid and pure shunt active filters, in *Conference Record IEEE-IAS Annual Meeting*, vol. 2, 2003, pp. 1195–1202
29. S. Srianthumrong, H. Akagi, A medium-voltage transformerless AC/DC Power conversion system consisting of a diode rectifier and a shunt hybrid filter. IEEE Trans. Indus. Appl. **39**, 874–882 (2003)
30. W. Tangtheerajaroonwong, T. Hatada, K. Wada, H. Akagi, Design and performance of a transformerless shunt hybrid filter integrated into a three-phase diode rectifier. IEEE Trans. Power Electron. **22**, 1882–1889 (2007)
31. H.-L. Jou, K.-D. Wu, J.-C. Wu, C.-H. Li, M.-S. Huang, Novel power converter topology for three phase four-wire hybrid power filter. IET Power Electron. **1**, 164–173 (2008)
32. R. Inzunza, H. Akagi, A 6.6-kV transformerless shunt hybrid active filter for installation on a power distribution system. IEEE Trans. Power Electron. **20**, 893–900 (2005)
33. V.F. Corasaniti, M.B. Barbieri, P.L. Arnera, M.I. Valla, Hybrid power filter to enhance power quality in a medium voltage distribution. IEEE Trans. Indus. Electron. **56**, 2885–2893 (2009)
34. S. Rahmani, A. Hamadi, N. Mendalek, K. Al-Haddad, A new control technique for three-phase shunt hybrid power filter. IEEE Trans. Indus. Electron. **56**, 2904–2915 (2009)
35. W.R.A. Ibrahim, M.M. Morcos, Artificial intelligence and advanced mathematical tools for power quality applications: a survey. IEEE Trans. Power Del. **17**(2)

Chapter 2
Basic Principles for Parallel Power Electronic Filters

Abstract In this chapter, the basic principles for parallel power electronics filters are discussed. In Sect. 2.1, the influences of reactive power, harmonics and unbalance are given, which affect and degrade the practical power network. The basic circuit configuration for injecting and absorbing compensating current is discussed. Furthermore, several parallel power filters are given and compared. In Sect. 2.2, instantaneous power theories which are important for detecting the compensation current components and controlling the fast response of power electronic compensators are introduced and addressed. Finally, different power electronic converters are discussed and compared in Sect. 2.3. The effect of dc voltage variation on the switching functions are also introduced in Sect. 2.3.

2.1 Parallel Compensation for Reactive Power, Current Harmonics and Unbalance

In an ideal case, to reduce the oscillating power components flowing between the source and loads, the voltage and current should be sinusoidal, in-phase and balanced. Thus, the electric system can be operated with higher efficiency and better power quality without reactive power, harmonics and unbalance problems in the power network. However, in practical cases, due to the common loads which are normally inductive and non-linear as well as the unpredictable loadings in different phases in a three-phase system, there will be reactive power oscillating around the system, and distorted unbalance current waveforms would exist. Reactive power, current harmonics and unbalance problems usually occur in practical power networks, in which the influences of reactive power, harmonics and unbalance are given as follows. After that, the basic principle of parallel compensators is introduced in this section.

© Springer Science+Business Media Singapore 2016
M.-C. Wong et al., *Parallel Power Electronics Filters in Three-Phase Four-Wire Systems*, DOI 10.1007/978-981-10-1530-4_2

2.1.1 Influence of Reactive Power

The reactive power occurs when there is a phase angle difference between the voltage and current under a storage passive load. The reactive power in an electric network causes the following issues:

(1) The reactive power increases the system capacity rating. Increasing the reactive power causes the source current to increase. As a result, the apparent power is increased too. Thus this increases system capacity ratings of generators, transformers, transmission lines, electric applications and measurement systems. Finally, all initial costs are increased proportionally.

(2) The reactive power is an oscillating power between the source and the load; it increases the unnecessary source current. As a result, the transmission loss is increased by the reactive power flowing inside the system. Under a steady-state condition, the reactive power is expressed as Q. The root mean square (RMS) source current can be decomposed into active current, I_p, and reactive current, I_q. The transmission loss can be represented by (2.1), where R is the transmission line total resistance and V is the RMS source voltage.

$$\Delta P = I^2 R = \left(I_p^2 + I_q^2 \right) R = \frac{P^2 + Q^2}{V^2} R \qquad (2.1)$$

The second term of (2.1), $(Q^2/V^2)R$ causes the additional power loss by reactive power flow. When all the instantaneous reactive power can be compensated, the transmission loss can be reduced.

(3) The voltage variation of transformers and feeding networks can be increased. Figure 2.1 shows an equivalent circuit for reactive power influence on transformers, where V_s is the source voltage, a transmission line normally is inductive with R (transmission line resistance) and X (transmission line inductance), the feeding terminal voltage V with equivalent loading admittance for transformers such as G (conductance) and B (inductive admittance) and system current I.

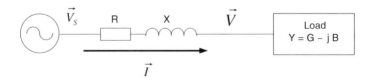

Fig. 2.1 Equivalent circuit for reactive power influence on transformers

Fig. 2.2 Phasor diagram for reactive power influence on transformers

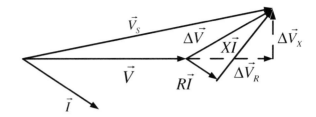

From Fig. 2.2, the phasor diagram shows that the transmission line impedance causes the terminal voltage drop, ΔV. $\Delta \vec{V}$ means a terminal voltage vector that can be decomposed into a real part and an imaginary part respectively.

$$\Delta \vec{V} = \vec{V}_S - \vec{V} = (R + jX) \cdot \vec{I} \tag{2.2}$$

$$\vec{I} = V(G - jB) = \frac{V^2 G - jV^2 B}{V} = \frac{P - jQ}{V} \tag{2.3}$$

$$\Delta \vec{V} = (R + jX) \cdot \left(\frac{P - jQ}{V}\right) = \frac{RP + XQ}{V} + j\frac{XP - RQ}{V} = \Delta V_R + j\Delta V_X \tag{2.4}$$

By checking the phasor diagram, due to the small phase angle difference between $\vec{V}_S$ and $\vec{V}$, it shows that $\Delta V \approx \Delta V_R$. Finally,

$$\Delta V \approx \Delta V_R = \frac{RP + XQ}{V} \tag{2.5}$$

Normally, X >> R. As a result, the active power does not influence too much on the terminal voltage variation. On the other hand, the reactive power causes more terminal voltage variation. Voltage variation can affect the normal operation of other electric applicants and finally the quality of power supplied to the users.

2.1.2   Influence of Harmonics

In the ideal case, the voltage and current should be sinusoidal, which do not have any harmonic component. A harmonic of a wave is a component frequency of a signal that is an integer multiple of the fundamental frequency, where the fundamental frequency is the system operating frequency which is 50 Hz in China and 60 Hz in America. In practical cases, the voltage and current cannot or may not be sinusoidal. Figure 2.3 shows practical measured voltage and current waveforms respectively, it shows that the current is highly distorted and the voltage quality is affected due to the current distortion.

Table 2.1 depicts the harmonic amplitudes of the current waveform of Fig. 2.3, which shows that its Total Harmonic Distortion (THD) is 127.39 %.

Fig. 2.3 Practical voltage
and current waveforms

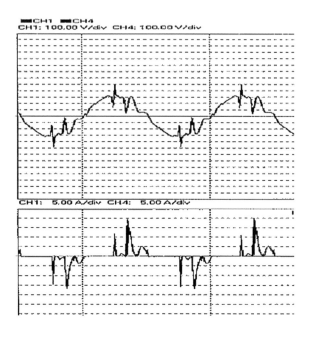

Table 2.1 Harmonic amplitude of current waveforms in Fig. 2.3

CH1: Value harmonic								
Order		(A)	Order		(A)	Order		(A)
1	100.00	0.48	18	21.98	0.74	35	4.19	0.48
2	5.47	0.49	19	16.13	0.43	36	8.51	0.49
3	52.17	0.40	20	22.16	0.56	37	6.36	0.55
4	5.70	0.43	21	8.54	0.29	38	5.89	0.39
5	24.24	0.30	22	9.87	0.31	39	8.65	0.50
6	15.58	0.41	23	4.93	0.28	40	9.24	0.30
7	44.01	0.33	24	2.28	0.37	41	7.53	0.42
8	24.43	0.41	25	8.27	0.42	42	7.97	0.31
9	59.72	0.35	26	9.49	0.47	43	7.53	0.46
10	23.15	0.35	27	5.39	0.51	44	6.35	0.34
11	40.38	0.22	28	13.71	0.52	45	2.84	0.41
12	14.66	0.33	29	3.62	0.36	46	3.52	0.31
13	17.03	0.30	30	12.95	0.45	47	4.45	0.37
14	11.81	0.56	31	5.79	0.32	48	4.13	0.37
15	14.71	0.44	32	9.54	0.35	49	9.14	0.39
16	16.02	0.68	33	5.64	0.37	50	0.33	0.44
17	19.74	0.49	34	3.28	0.35	THD	127.39	3.00 (%)

The harmonics existed in power networks can cause the following problems:

(1) Harmonic current can cause additional power loss and reduce the system efficiency of generators, power transmission and electric utilities. When large amounts of triple harmonic current exist in the neutral wire in a three-phase four-wire system, it may cause overheating and even a fire at the neutral wire.
(2) Harmonics influence normal operations of electric applicants and can reduce the expected life period of electric applicants.
(3) Harmonics may cause series and parallel resonance in a specific region of the power networks.
(4) Harmonics may trigger the protection system which causes power breakdown or operates abnormally.
(5) The power measurement system may include calculation errors.
(6) The communication system can be influenced by harmonics due to electro-magnetic interference.

2.1.3 Influence of Unbalance

In an ideal balance three-phase system, the three-phase voltage RMS amplitudes should be the same and phase angle difference among them should be 120° apart from each other. However, in practical cases, the voltage and current may not be balanced. Figure 2.4 exhibits the measured three-phase current and the neutral current waveforms in a distributed transformer, which shows that the system is operated under unbalance and with high neutral current amplitude.

The unbalance exists in power networks can cause the following problems:

(1) In low voltage distribution sites, three-phase unbalance loading may cause user's voltage variation. On one side, the terminal voltage may be higher and it may destroy electric applications, and the electric applicants in low voltage side may not operate normally. Figure 2.5 shows a three-phase four-wire

Fig. 2.4 Three phase current and neutral current waveforms

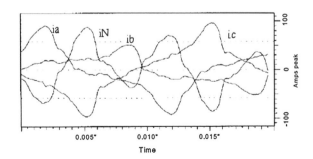

Fig. 2.5 A three-phase
four-wire system

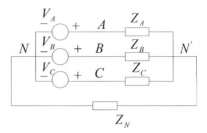

Fig. 2.6 A three-phase
four-wire system unbalance
phase diagram

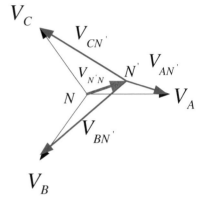

system connected with different load impedances, Z_A, Z_B and Z_C.
Equation (2.6) shows that the neutral voltage $V_{N'N}$ may not be zero if the loads
are unbalance with balanced voltage situation, in which Y_A, Y_B and Y_C are
equivalent loading admittance Fig. 2.6 displays the phase diagram for terminal
user voltages, $V_{AN'}$, $V_{BN'}$ and $V_{CN'}$, which shows that the terminal user voltages
can be varied if the loads are unbalanced and the neutral line impedance, Z_N, is
not small enough.

$$V_{N'N} = \frac{V_A Y_A + V_B Y_B + V_C Y_C}{Y_A + Y_B + Y_C + \frac{1}{Z_N}} \tag{2.6}$$

(2) There is an additional loss in the electric motors and generators due to the
     negative sequence of the unbalance current, and negative sequence can reduce
     the resultant torque of a motor and its efficiency, and finally it causes the
     vibration between the stator and rotor of a motor.
(3) Abnormal operation of protected systems in a power network degrades the
     stability and security of the power system.
(4) Unbalance voltage may cause additional harmonics from the semiconductor
     converters.

(5) The system capacity utilization ratio of a generator can be degraded.
(6) The operation point of a transformer that may shift up or down causes harmonic distortion on the voltage waveforms and additional loss inside the transformer, which increases the temperature of the transformer and its peripheral environment.
(7) The unbalance operation degrades the electric system efficiency and the quality of telecommunication systems.
(8) High neutral current causes overheating and failure of the neutral conductor.

2.1.4 Basic Principle of Parallel Compensation

Figure 2.7 shows the basic configuration for parallel current quality compensation. Figure 2.8 shows its corresponding waveforms for compensation. The compensator is connected in parallel with the load. It shows a source supplying a non-linear load that is being compensated by a shunt compensator. As a result, the source current, $\vec{i}_S$ can be balanced and sinusoidal. However, the load current, $\vec{i}_L$ does not change before and after compensation. The load currents of non-linear unbalanced loads are still unbalance, non-sinusoidal and with harmonics. On the other hand, the compensator injected current, $\vec{i}_C$ turns the source current to be balanced and sinusoidal. The ideal compensator is a controlled current source that can inject or draw any set of arbitrarily compensating current reference, $\vec{i}_C^*$.

Basically, there are 4 different kinds of parallel compensators: Capacitor Banks (CB), Passive Power Filters (PPF), Active Power Filters (APF) and Hybrid Active Power Filters (HAPP).

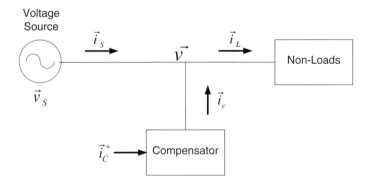

Fig. 2.7 Basic configuration for parallel compensation

Fig. 2.8 Corresponding
compensation waveforms

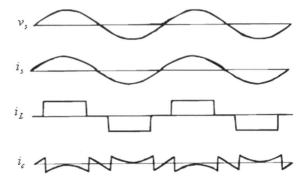

Fig. 2.9 Parallel capacitor
banks

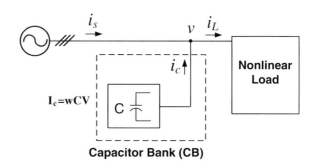

2.1.4.1 Capacitor Banks (CB)

Shunt Connected Capacitor Banks (CB) are extensively used in distribution power
networks for power factor correction and feeder voltage control due to its relatively
low cost and simplicity, which are developed mainly for reactive power compensation as shown in Fig. 2.9. However, since the current passing through the
capacitor is proportional to the frequency and rate of change of voltage, the parallel
capacitor banks may possibly be damaged and may even cause explosion as shown
in Fig. 2.10 if the harmonics level is high. Moreover, CBs have other disadvantages
such as static compensation capability (low dynamic performance), slow response
time and resonance phenomena.

2.1.4.2 Passive Power Filters (PPF)

The first installation of Passive Power Filters (PPF) was in 1940s [1], PPFs have
been widely used to suppress current harmonics and to compensate reactive power
in distribution power systems due to its low cost, simplicity and high efficiency [2–
4]. The system configurations of conventional passive power filters (PPF) are
shown in Fig. 2.11. PPFs have been considered a good alternative for current
harmonic compensation and displacement power factor correction. In general,

Fig. 2.10 Capacitor bank explosion in a power sub-station

Fig. 2.11 Parallel passive power filters

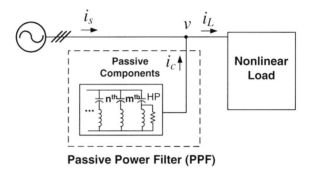

passive tuned filters have been used to eliminate low frequency current harmonics while high-pass passive units have been connected to attenuate the amplitude of high frequency current components [5, 6].

Technical disadvantages of PPFs have been extensively discussed and the inherent drawbacks [2–7] are:

- Source impedance strongly affects filtering characteristics.
- PPF may fall in series resonance when source harmonic voltage produces excessive harmonic current flowing into PPF.
- The shunt PPF may fall in parallel resonance with the source impedance.
- Small design tolerances are acceptable in the values of L and C. Small variations in the values of L or C modify the filter resonant frequency.

In addition, the PPF cannot modify their compensation characteristics according to the dynamic changes of the nonlinear load. The PPF generates fundamental frequency reactive power that changes the system voltage regulation, and if the filter is not designed properly or disconnected during low load operating conditions, over-voltage can be generated at its terminals [8].

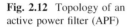

Fig. 2.12 Topology of an active power filter (APF)

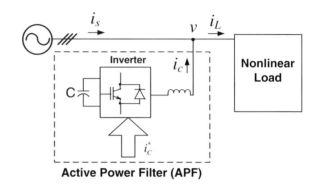

Active Power Filter (APF)

2.1.4.3 Active Power Filters (APF)

Since the concept "Active ac Power Filter" was first developed by Gyugyi in 1976 [9], the research studies of the active power filters (APFs) for current quality compensation prosper. With the remarkable progress in the speed and capacity of semiconductors switching devices, such as GTO thyristors and IGBTs, active power filters consisting of voltage or current source pulse width modulation (PWM) inverters have been studied and put into practical use [9, 10]. The APFs improve the power quality by injecting equal but opposite currents to compensate harmonic distortion, reactive power and unbalance currents in the power network. Figure 2.12 shows the system configuration of an APF. APFs consisting of voltage-fed PWM inverters using IGBTs or GTO thyristors are operating successfully, making significant contributions to improving power quality [11]. The APFs have the ability to overcome the above-mentioned disadvantages inherent in PPFs. However, the initial costs and running costs of the APFs are much higher than those of the PPFs [11, 12] due to the costs of semiconductor switching devices with its drivers, signal conditioning boards and digital controller(s).

2.1.4.4 Hybrid Active Power Filters (HAPF)

The CBs and PPFs cannot achieve dynamic reactive power and current quality compensation but their costs are relatively low. Sometimes, they may cause the operation problems in the power networks and even explosion. The APFs can overcome the above issues, but require high initial and operational costs. There is an alternative that is to combine both PPF and APF to form a Hybrid Active Power Filter (HAPF) [2, 11]. Hybrid active power filters (HAPFs) consist of active and passive filters connected in series or parallel to each other so that the advantages of both filters can be combined. The main purpose of the HAPF is to improve the compensation characteristics of PPF and reduce the costs of the APF. It leads to the effective balance in cost and performance for current quality compensation. Figure 2.13 shows a typical topology of an HAPF.

Fig. 2.13 A typical topology
of a hybrid active power filter
(HAPF)

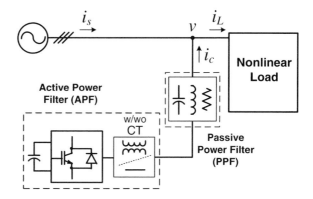

In this book, the APF and HAPF are the focus. More details are discussed in Chaps. 3 and 4. In next section, the instantaneous compensation current technique to control the voltage source inverter (VSI) so as to inject or absorb the required compensating current instantaneously will be discussed in detail.

2.2   Instantaneous Power Compensation

Conventionally, the active, reactive and apparent powers are defined under steady-state, which are calculated by root-mean-square (RMS) voltage and current values and/or phase angle difference between the voltage and the current. It means that the compensation or the measurement equipment needs to wait at least one period long, for example 0.02 s in a 50 Hz frequency system, in order to get power components for further operations. As a result, the speed for the response by the compensators is relatively low.

In 1983 [13, 14], the instantaneous active and reactive power theory or the p-q theory was first introduced by Akagi et al. in Japan. The p-q theory [14, 15] pioneered in transforming voltages and currents from abc to αβ coordinates. This concept has provided an effective method to compensate the instantaneous components of reactive power for the three-phase system without energy storage. However, this instantaneous reactive power theory has a conceptual limitation as pointed out in [16], i.e., the theory is only complete for the three-phase systems without zero-sequence currents and voltages.

To resolve this limitation and associated problems, Willems and Nabae proposed some attractive approaches to define the instantaneous active and reactive currents [16, 17]. However, their approaches are to deal with the decomposition of currents into orthogonal components rather than with the power components. In 1996, a generalized theory of instantaneous reactive power has been proposed [18–21]. The generalized theory is valid for sinusoidal or non-sinusoidal and balanced or unbalanced three-phase systems, with or without zero-sequence currents and/or voltages.

In this part, the definitions of this generalized instantaneous reactive power theory are given both in *a-b-c* and *α-β-0* coordinates. Finally, the compensating current can be calculated and is discussed in this section.

The instantaneous active power p and reactive power q of a three-phase system are defined as:

$$p = \vec{v} \cdot \vec{i} \tag{2.7}$$

$$\vec{q} = \vec{v} \times \vec{i} \tag{2.8}$$

$$q = \|\vec{q}\| \tag{2.9}$$

where "·" denotes the dot (internal) product or scalar product of vectors, "×" denotes the cross (exterior) product of vectors or vector product, *v* and *i* are the voltage and current vectors. Since the instantaneous reactive power obtained from Eq. (2.8) is a vector, the magnitude or the length of the instantaneous reactive power is often used, which is defined in Eq. (2.9).

2.2.1 Definitions in a-b-c Frame

For a three-phase system, instantaneous voltages v_a, v_b, v_c and instantaneous currents i_a, i_b, i_c are expressed as instantaneous space vector $\vec{v}$ and $\vec{i}$, i.e.,

$$\vec{v} = v_a \vec{n}_a + v_b \vec{n}_b + v_c \vec{n}_c \tag{2.10}$$

$$\vec{i} = i_a \vec{n}_a + i_b \vec{n}_b + i_c \vec{n}_c \tag{2.11}$$

where $(\vec{n}_a, \vec{n}_b, \vec{n}_c)$ is a basis in *a-b-c* frame.

By using the definitions of (2.7) and (2.8) the instantaneous active and reactive powers can be expressed as:

$$p = \vec{v} \cdot \vec{i} = v_a i_a + v_b i_b + v_c i_c \tag{2.12}$$

$$\vec{q} = \vec{v} \times \vec{i} = q_a \vec{n}_a + q_b \vec{n}_b + q_c \vec{n}_c \tag{2.13}$$

where

$$q_a = \begin{vmatrix} v_b & i_b \\ v_c & i_c \end{vmatrix}, \quad q_b = \begin{vmatrix} v_c & i_c \\ v_a & i_a \end{vmatrix}, \quad q_c = \begin{vmatrix} v_a & i_a \\ v_b & i_b \end{vmatrix}$$

Finally, instantaneous reactive power is:

$$\vec{q} = (v_b i_c - v_c i_b)\vec{n}_a + (v_c i_a - v_a i_c)\vec{n}_b + (v_a i_b - v_b i_a)\vec{n}_c \quad (2.14)$$

The instantaneous active current vector $\vec{i}_p$ and the instantaneous reactive current vector $\vec{i}_q$ are expressed as:

$$\vec{i}_p = \begin{bmatrix} i_{ap} \\ i_{bp} \\ i_{cp} \end{bmatrix} = \frac{p}{\vec{v} \cdot \vec{v}}\vec{v} = \frac{p}{v^2}\vec{v} \quad (2.15)$$

$$\vec{i}_q = \begin{bmatrix} i_{ap} \\ i_{bp} \\ i_{cp} \end{bmatrix} = \frac{\vec{q} \times \vec{v}}{\vec{v} \cdot \vec{v}} = \frac{\vec{q} \times \vec{v}}{v^2} \quad (2.16)$$

where $v = \|\vec{v}\| = \sqrt{v_a^2 + v_b^2 + v_c^2}$ and $\vec{i} = \vec{i}_p + \vec{i}_q$

2.2.2 Definitions in α-β-0 Frame

The α-β-0 transformation of a three-phase four-wire system gives:

$$\begin{bmatrix} i_0 \\ i_\alpha \\ i_\beta \end{bmatrix} = [P] \cdot \begin{bmatrix} i_a \\ i_b \\ i_c \end{bmatrix} \quad (2.17)$$

$$\begin{bmatrix} v_0 \\ v_\alpha \\ v_\beta \end{bmatrix} = [P] \cdot \begin{bmatrix} v_a \\ v_b \\ v_c \end{bmatrix} \quad (2.18)$$

where

$$P = \sqrt{\frac{2}{3}}\begin{bmatrix} 1/\sqrt{2} & 1/\sqrt{2} & 1/\sqrt{2} \\ \cos\theta & \cos(\theta - 120°) & \cos(\theta + 120°) \\ -\sin\theta & -\sin(\theta - 120°) & -\sin(\theta + 120°) \end{bmatrix}$$

$$= \sqrt{\frac{2}{3}}\begin{bmatrix} 1/\sqrt{2} & 1/\sqrt{2} & 1/\sqrt{2} \\ 1 & -1/2 & -1/2 \\ 0 & \sqrt{3}/2 & -\sqrt{3}/2 \end{bmatrix}$$

for $\theta = 0$, then $[P]^T = [P]^{-1}$, the matrix $[P]$ is the Clarke Transformation which transfers voltages and currents from a-b-c to α-β-0 coordinates.

As a result, the instantaneous active power can be expressed as:

$$p_{(\alpha\beta0)} = \vec{v} \cdot \vec{i} = \vec{v}_{(\alpha\beta0)} \cdot \vec{i}_{(\alpha\beta0)} = v_\alpha i_\alpha + v_\beta i_\beta + v_0 i_0 \qquad (2.19)$$

where the suffixes "(abc)" and "$(\alpha\beta0)$" denote the corresponding coordinates. Then, its instantaneous active power can be expressed as:

$$p_{(abc)} = \vec{v}_{(abc)} \cdot \vec{i}_{(abc)} = [\vec{v}]^T [\vec{i}]$$
$$p_{(\alpha\beta0)} = \vec{v}_{(\alpha\beta0)} \cdot \vec{i}_{(\alpha\beta0)} = ([P][\vec{v}])^T [P][\vec{i}] = [\vec{v}]^T [P]^T [P][\vec{i}]$$
$$= [\vec{v}]^T [I][\vec{i}] = [\vec{v}]^T [\vec{i}]$$

so $p_{(\alpha\beta0)} = \vec{v} \cdot \vec{i} = \vec{v}_{(\alpha\beta0)} \cdot \vec{i}_{(\alpha\beta0)} = \vec{v}_{(abc)} \cdot \vec{i}_{(abc)} = p_{(abc)}$

Furthermore, the instantaneous reactive power can be expressed as:

$$\vec{q}_{(\alpha\beta0)} = \vec{v}_{(\alpha\beta0)} \times \vec{i}_{(\alpha\beta0)} = (v_\beta i_0 - v_0 i_\beta)\vec{n}_\alpha + (v_0 i_\alpha - v_\alpha i_0)\vec{n}_\beta + (v_\alpha i_\beta - v_\beta i_\alpha)\vec{n}_0 \qquad (2.20)$$

$$\vec{q}_{(\alpha\beta0)} = \vec{v}_{(\alpha\beta0)} \times \vec{i}_{(\alpha\beta0)} = ([P]\vec{v}_{(abc)}) \times ([P]\vec{i}_{(abc)})$$
$$= [P](\vec{v}_{(abc)} \times \vec{i}_{(abc)}) = [P]\vec{q}_{(abc)}$$

because $\|[P]\| = 1$ and $\|q_{(abc)}\| = \|q_{(\alpha\beta0)}\|$ can be deduced.

The previous discussion indicates that the definitions of instantaneous active and reactive power can be expressed in both abc and $\alpha\beta0$ coordinates, and the used coordinates are linear independent to each other.

Similarly, the instantaneous active and reactive current components in α-β-0 coordinates can be written as:

$$\vec{i}_{p(\alpha\beta0)} = \frac{p}{\vec{v}_{(\alpha\beta0)} \cdot \vec{v}_{(\alpha\beta0)}} \vec{v}_{(\alpha\beta0)} \qquad (2.21)$$

$$\vec{i}_{q(\alpha\beta0)} = \frac{\vec{q} \times \vec{v}_{(\alpha\beta0)}}{\vec{v}_{(\alpha\beta0)} \cdot \vec{v}_{(\alpha\beta0)}} \qquad (2.22)$$

In addition, if the generalized instantaneous reactive power theory is applied to the three-phase three-wire system, i.e., $v_0 = 0$ and $i_0 = 0$, the following results can be derived.

$$p = \vec{v}_{(\alpha\beta0)} \cdot \vec{i}_{(\alpha\beta0)} = v_\alpha i_\alpha + v_\beta i_\beta \qquad (2.23)$$

$$q = \|\vec{v}_{(\alpha\beta0)} \times \vec{i}_{(\alpha\beta0)}\| = v_\alpha i_\beta - v_\beta i_\alpha \qquad (2.24)$$

2.2.3 Mixed-Coordinate Instantaneous Compensation

Based on previous parts, three-phase voltages and currents can be described by the 3-dimensional base either in *abc* or *αβ0* coordinates. In general, the equivalent voltage and current vectors can be expressed by (2.25) and (2.26) in xyz coordinate. The xyz can be abc or αβ0 coordinates

$$\vec{v}_e = v_x \vec{n}_x + v_y \vec{n}_y + v_z \vec{n}_z \tag{2.25}$$

$$\vec{i}_e = i_x \vec{n}_x + i_y \vec{n}_y + i_z \vec{n}_z \tag{2.26}$$

The instantaneous active and reactive powers can be expressed by (2.27) and (2.28), which can be further decomposed into average and oscillating components as given in (2.29) and (2.30).

$$p = \vec{v}_e \cdot \vec{i}_e = v_x i_x + v_y i_y + v_z i_z \tag{2.27}$$

$$\vec{q} = \vec{v}_e \times \vec{i}_e = (v_x i_y - v_y i_x)\vec{n}_z + (v_y i_z - v_z i_y)\vec{n}_x + (v_z i_x - v_x i_z)\vec{n}_y \tag{2.28}$$

$$p = \bar{p} + \tilde{p} \tag{2.29}$$

$$\vec{q} = \bar{q} + \tilde{q} \tag{2.30}$$

Additionally, the instantaneous active and reactive current can be expressed as follows:

$$\vec{i}_p = \frac{p \cdot \vec{v}_e}{\vec{v}_e \cdot \vec{v}_e} \tag{2.31}$$

$$\vec{i}_q = \frac{q \times \vec{v}_e}{\vec{v}_e \cdot \vec{v}_e} \tag{2.32}$$

Finally, its injecting current for compensation can be expressed by (2.33), which shows that only oscillating component of instantaneous active power and all instantaneous reactive power are compensated.

$$\vec{i}_c^* = \frac{\tilde{p} \cdot \vec{v}_e}{\vec{v}_e \cdot \vec{v}_e} + \frac{\vec{q} \times \vec{v}_e}{\vec{v}_e \cdot \vec{v}_e} \tag{2.33}$$

No matter the voltages or currents are expressed in abc or αβ0 coordinates, they are describing the same voltage or current vectors in 3-dimensional base. One should be noted that the voltage sources should be assumed to be balanced and sinusoidal. Otherwise, the instantaneous compensated current in (2.33) should be further modified as there are negative and/or zero sequence average power components existed in instantaneous average active power. Without further modification, the compensated source currents are not sinusoidal and not balanced for unbalanced and/or non-sinusoidal voltages, in which further discussions for unbalanced and non-sinusoidal voltage cases are not included in this book.

The instantaneous compensating currents can be calculated by mixed coordinate signals. As a result, the computation steps and time by a digital controller, such as digital signal processor (DSP) can be reduced. The computation error can also be lowered, in which the detailed algorithm description immediately follows.

The instantaneous compensating active current can be expressed as:

$$\vec{i}_{cp} = \frac{\tilde{p} \cdot \vec{v}_e}{\vec{v}_e \cdot \vec{v}_e} \tag{2.34}$$

where $\tilde{p}$ is the oscillating part of the instantaneous active power which can be calculated by

$$p = v_a i_a + v_b i_b + v_c i_c = p_a + p_b + p_c \tag{2.35}$$

The dot product of the instantaneous voltage is

$$\Delta = \vec{v}_e \cdot \vec{v}_e = v_a^2 + v_b^2 + v_c^2 \tag{2.36}$$

Finally, its instantaneous compensating active current can be expressed as:

$$\vec{i}_{cp} = \frac{\tilde{p}}{\Delta} \left(v_\alpha \vec{n}_\alpha + v_\beta \vec{n}_\beta + v_0 \vec{n}_0 \right) = \frac{\tilde{p}}{\Delta} \left(v_a \vec{n}_a + v_b \vec{n}_b + v_c \vec{n}_c \right) \tag{2.37}$$

All the instantaneous reactive power components are required for calculating the instantaneous reactive compensating current,

$$\vec{i}_{cq} = \frac{q \times \vec{v}_e}{\vec{v}_e \cdot \vec{v}_e} \tag{2.38}$$

The instantaneous reactive compensating power vector components can be projected on $\{\vec{n}_a, \vec{n}_b, \vec{n}_c\}$ base. The instantaneous reactive power can be calculated by (2.39). The cross product of reactive power $\vec{q}$ and voltage vector $\vec{v}_e$ can be derived from (2.40).

$$\vec{q} = (v_b i_c - v_c i_b)\vec{n}_a + (v_c i_a - v_a i_c)\vec{n}_b + (v_a i_b - v_b i_a)\vec{n}_c = A\vec{n}_a + B\vec{n}_b + C\vec{n}_c \tag{2.39}$$

$$\vec{q} \times \vec{v}_e = \begin{vmatrix} \vec{n}_a & \vec{n}_b & \vec{n}_c \\ A & B & C \\ v_a & v_b & v_c \end{vmatrix} = (Bv_c - Cv_b)\vec{n}_a + (Cv_a - Av_c)\vec{n}_b + (Av_b - Bv_a)\vec{n}_c$$

$$\tag{2.40}$$

where

$$A = (v_b i_c - v_c i_b), B = (v_c i_a - v_a i_c) \text{ and } C = (v_a i_b - v_b i_a)$$

Finally, substitute (2.40) into (2.38), its instantaneous compensating reactive current can be described by (2.41).

$$\vec{i}_{cq} = \frac{1}{\Delta} \left[\left(v_c^2 i_a + v_b^2 i_a - v_a v_c i_c - v_a v_b i_b \right) \vec{n}_a \right.$$
$$+ \left(v_a^2 i_b + v_c^2 i_b - v_b v_a i_a - v_b v_c i_c \right) \vec{n}_b$$
$$\left. + \left(v_b^2 i_c + v_a^2 i_c - v_c v_b i_b - v_c v_a i_a \right) \vec{n}_c \right] \tag{2.41}$$

By just taking out the first term of (2.41) and substituting (2.35) into it, after simplification, the instantaneous compensating reactive current in phase a can be obtained as given in (2.42).

$$\vec{i}_{cqa} = \frac{1}{\Delta} \left(v_c^2 i_a + v_b^2 i_a - v_a v_c i_c - v_a v_b i_b \right)$$
$$= \frac{1}{\Delta} \left(i_a (v_b^2 + v_c^2) - v_a p_b - v_a p_c \right)$$
$$= \frac{1}{\Delta} \left(i_a (\Delta - v_a^2) - v_a p_b - v_a p_c \right) \tag{2.42}$$
$$= \frac{1}{\Delta} \left(i_a \Delta - v_a p_a - v_a p_b - v_a p_c \right)$$
$$= i_a - \frac{p}{\Delta} v_a$$

By considering phase b and phase c as well, the instantaneous compensating reactive current in 3-dimensional case can be given in (2.43).

$$\vec{i}_{cq} = \left(i_a - \frac{p}{\Delta} v_a \right) \vec{n}_a + \left(i_b - \frac{p}{\Delta} v_b \right) \vec{n}_b + \left(i_c - \frac{p}{\Delta} v_c \right) \vec{n}_c \tag{2.43}$$

The final instantaneous compensating current is composed of instantaneous compensating active and reactive currents as shown in (2.44). After simplification, the final instantaneous compensating current for abc coordinates is shown in (2.45).

$$\vec{i}_c^* = \vec{i}_{cp} + \vec{i}_{cq} \tag{2.44}$$

$$\vec{i}_c^* = \left(i_a - \frac{\bar{p}}{\Delta} v_a \right) \vec{n}_a + \left(i_b - \frac{\bar{p}}{\Delta} v_b \right) \vec{n}_b + \left(i_c - \frac{\bar{p}}{\Delta} v_c \right) \vec{n}_c \tag{2.45}$$

Moreover, the voltage and current vectors can also be expressed in $\alpha\beta0$ coordinates; its compensating current can be described by (2.46).

$$\vec{i}_c^* = \left(i_\alpha - \frac{\bar{p}}{\Delta} v_\alpha\right)\vec{n}_\alpha + \left(i_\beta - \frac{\bar{p}}{\Delta} v_\beta\right)\vec{n}_\beta + \left(i_0 - \frac{\bar{p}}{\Delta} v_0\right)\vec{n}_0 \qquad (2.46)$$

where $\Delta = v_a^2 + v_b^2 + v_c^2$, $p = v_a i_a + v_b i_b + v_c i_c$ and $\bar{p}$ is the average value of p. Figure 2.14 shows the mixed coordinate instantaneous compensation algorithm.

Simulation verifications are performed by Matlab/Simulink. Figure 2.15 shows the three-phase source voltages while Fig. 2.16 shows the three-phase currents, the currents are obviously unbalanced and with harmonics. Figure 2.17 shows the source currents after transformation in α, β and 0 coordinates.

Figure 2.18 shows the compensated source currents with the compensation consideration of instantaneous reactive current only. It clearly shows that the source currents are not perfectly sinusoidal and balanced after compensation.

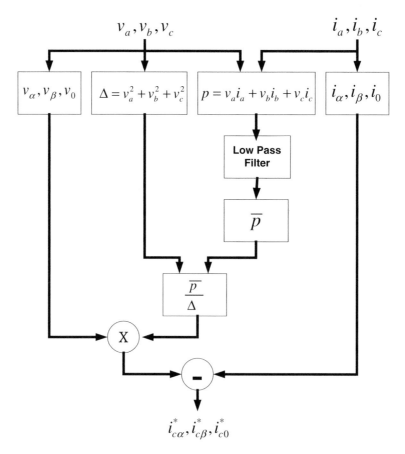

Fig. 2.14 Mixed coordinate instantaneous compensation signal flow diagram

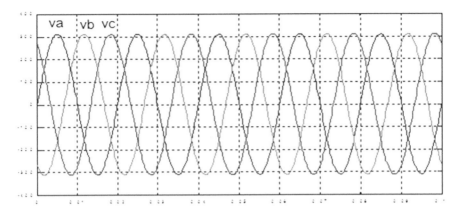

Fig. 2.15   Three-phase source voltage

Fig. 2.16   Three-phase
currents

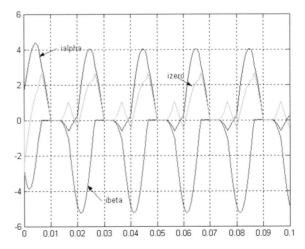

Fig. 2.17   Three-phase
source currents in $\alpha\beta 0$
coordinates

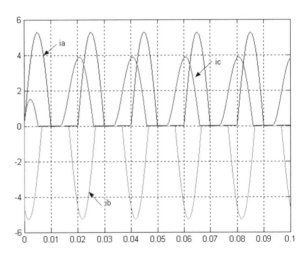

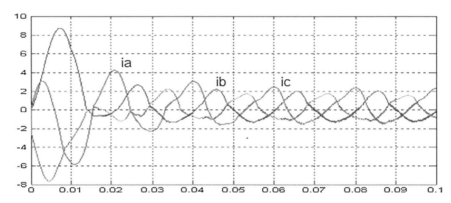

Fig. 2.18 Compensated source currents with compensation consideration of instantaneous reactive current only

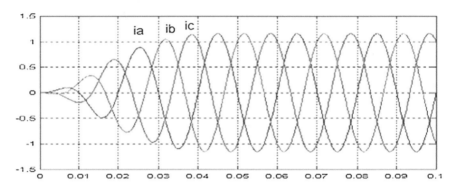

Fig. 2.19 Compensated source currents by mixed coordinate instantaneous compensation algorithm

Figure 2.19 shows the simulated result by Mixed Coordinate Instantaneous Compensation Algorithm. It clearly shows that three-phase source currents are sinusoidal and balanced after compensation. Figure 2.20 shows the α, β and 0 coordinate source currents after compensation by Mixed Coordinate Instantaneous Compensation Algorithm.

From the above simulation results, there is a transient part between $t = 0$ s and $t = 0.03$ s. Under the transient period, there is a large influence due to the transient response of the low-pass filter to obtain the instantaneous average power. The response of the low-pass filter affects this transient response. The faster the response of the low-pass filter, the shorter the transient period will be. However, the low-pass filter design is out of the scope of this section. The above mixed coordinate instantaneous power can simplify the calculation steps comparing to the original instantaneous reactive power theories.

Fig. 2.20 α, β and 0
coordinate source currents
after compensation

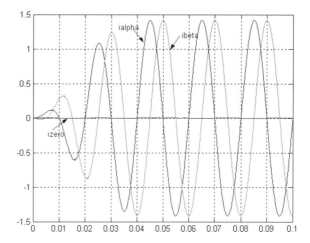

2.3   Three-Phase Converters and Their Discussions

In this section, three basic three-phase converter topologies are analyzed, compared and discussed in two-level and multi-level topologies such as three-leg, three-leg center-split and four-leg inverters in three-phase three-wire and three-phase four-wire systems. Since Year 2000, the 3-Dimensional Pulse Width Modulation (3D PWM) technique has become one of the popular research topics in three-phase four-wire applications, such as distributed power generators and active power filters [22–25]. The challenges to investigate the three-phase four-wire converters are:

(1)   To choose a suitable circuit topology,
(2)   To have a decoupling transformation mathematical model for analysis,
(3)   To develop a control strategy with high performance, and
(4)   To apply the system in high power applications.

The three-phase voltage source inverters (VSI) normally have two ways of providing a neutral connection for the three-phase four-wire systems: Three-Leg Center-Split and Four-Leg inverters [22]. However, according to the past studies [23, 25–27], most of the three-phase four-wire converter investigations mainly focus on the four-leg topology due to two reasons. First, the maximum achievable peak value of line-to-neutral voltage of a three-leg center-split inverter is lower than a four-leg inverter, which means that the utilization of dc linked voltage, V_{dc}, of a three-leg center-split inverter is lower. Second, large dc linked capacitors are needed in a three-leg center-split inverter to maintain a small ripple dc voltage. Based on the above points, the development of the control of the three-leg center-split topology is slower than that of the four-leg.

On the other hand, the control strategies of three-phase four-wire inverters are studied from abc Hysteresis controls to $\alpha\beta0$ space vector modulations [23–30] by abc-$\alpha\beta0$ transformations. Recently, there was 4×4 Quad-Transformation matrix

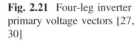

Fig. 2.21 Four-leg inverter
primary voltage vectors [27,
30]

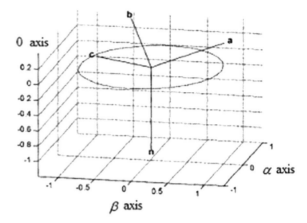

[27] which describes the direct transformation between the 4 Degree-of-Freedom
Leg-Output Voltage Space of the Four-Leg Inverter and its 3 Degree-of-Freedom
output voltage space. The Quad-Transform [27, 30] was proposed to analyze the
four-leg systems in $\alpha\beta0$ coordinates. The leg-voltages of a four-leg inverter could
be modeled as a projection into the 3 dimensional output-voltage space along 4
vectors through a Quad-transform. These vectors were denoted as the primary
voltage-vectors of the inverter and were depicted in Fig. 2.21. The a, b, c, and n
vectors in Fig. 2.21 formed the corners of a symmetrical tetrahedron [27, 30].

A unique 4 × 4 decoupling transformation matrix was presented for the four-leg
circuit topology, which was orthonormal, and hence invertible. The row and column vectors formed a basis for the 4 Degree-Of-Freedom leg-voltage space.
However, the concept of "Primary Voltage Vectors" that was defined as "An N-leg
inverter has an N-dimensional leg-voltage space, but only controls an (N − 1)
dimensional output voltage space. The leg voltages of an N-Leg inverter project
into the output-voltage space along the N vectors of equal magnitude equally
arranged in an (N − 1) Degree-Of-Freedom space [27, 30]". In this chapter, a 3
dimensional basis is used as follows. Furthermore, when four-leg and three-leg
center-split multilevel converters are taken into account in three-phase four-wire
high power applications, the advantages of the three-leg center-split multilevel
topology will be explained in the following sections.

2.3.1   Realization of 3 Dimensional Coordinates
for Three-Phase Systems

One should note that the voltage space vector development hereafter is based on the
assumption that instantaneous power is constant after and before transformation
from one into other coordinates in this book. In a three-phase three-wire or
three-phase four-wire system, an instantaneous voltage vector, $\vec{v}$, can be represented

by a linear combination of the vectors, $\vec{n}_a$, $\vec{n}_b$ and $\vec{n}_c$ which can form a basis, $B = \{\vec{n}_a, \vec{n}_b, \vec{n}_c\}$, for a vector to span with. The vectors, $\vec{n}_a$, $\vec{n}_b$ and $\vec{n}_c$ are linear independent to each other in a physical sense. Basically, the dimension of this instantaneous voltage vector is 3.

$$\vec{v} = K_a \vec{n}_a + K_b \vec{n}_b + K_c \vec{n}_c \tag{2.47}$$

where K_a, K_b and K_c are scalars.

Let $M = \{\vec{n}_1, \vec{n}_2 ... \vec{n}_n\}$ and $M' = \{\vec{w}_1, \vec{w}_2 ... \vec{w}_m\}$ be the two bases for a finite-dimensional vector space Q. Since M is a basis and M' is a linearly independent set, which implies that m = n. As a result, any two bases for a finite-dimensional vector space have the same number of vectors. An instantaneous voltage vector can be represented by $B = \{\vec{n}_a, \vec{n}_b, \vec{n}_c\}$ and $B' = \{\vec{n}_\alpha, \vec{n}_\beta, \vec{n}_0\}$, and it can span in a 3 dimension space. The 3-phase instantaneous voltages can be transferred from abc into $\alpha\beta0$ coordinates by the coordinate transformation matrix [19], $P_{\alpha\beta0}$, (2.48) under a linear power transformation consideration.

$$\begin{bmatrix} v_\alpha \\ v_\beta \\ v_0 \end{bmatrix} = \sqrt{\frac{2}{3}} \begin{bmatrix} 1 & -1/2 & -1/2 \\ 0 & \sqrt{3}/2 & -\sqrt{3}/2 \\ 1/\sqrt{2} & 1/\sqrt{2} & 1/\sqrt{2} \end{bmatrix} \begin{bmatrix} v_a \\ v_b \\ v_c \end{bmatrix} = P_{\alpha\beta0} \begin{bmatrix} v_a \\ v_b \\ v_c \end{bmatrix} \tag{2.48}$$

where $P_{\alpha\beta0} = \sqrt{\frac{2}{3}} \begin{bmatrix} \cos 0° & \cos(120°) & \cos(240°) \\ \sin 0° & \sin(120°) & \sin(240°) \\ \sin(45°) & \sin(45°) & \sin(45°) \end{bmatrix}$.

Under the linear power transformation consideration [19], there are two conditions that need to be fulfilled, (2.49) and (2.50).

$$I = [P_{\alpha\beta0}]^T [P_{\alpha\beta0}] \tag{2.49}$$

$$\left\| P_{\alpha\beta0} \right\| = 1 \tag{2.50}$$

From (2.49), it implies that $[P_{\alpha\beta0}]^{-1} = [P_{\alpha\beta0}]^T$, which has the property that the matrix $P_{\alpha\beta0}$ is said to be an orthogonal matrix and I is an identity matrix. As a result, vector coordinates can be described as shown in Fig. 2.22. Another important characteristic is noticed as follows. The determinant of $P_{\alpha\beta0}$ is equal to 1 so that the volumes bounded after and before the transformation are the same [31].

$$\left| P_{\alpha\beta0} \right| = 1 \tag{2.51}$$

It means that an instantaneous voltage vector can be expressed as (2.52).

$$\vec{v} = K_\alpha \vec{n}_\alpha + K_\beta \vec{n}_\beta + K_0 \vec{n}_0 \tag{2.52}$$

where K_α, K_β and K_0 are scalars.

Fig. 2.22 Transformation
basis allocation

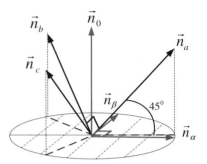

The zero sequence is the projection of vectors in a, b and c coordinates with 45°
on it. When the vector coordinates are considered from top view of Fig. 2.22, the
abc coordinates are apart from each other with 120°, but it should be noticed that
they are orthogonal to each other as shown in Fig. 2.22. All the transformations
taken by *abc* and *αβ0* coordinates are in 3 dimensions, not 3 to 2 dimensions in the
three-phase systems. It can also be explained why phase angles of *abc* coordinates
are 120° apart with orthogonal basis to each other. The above concept can be
applied in multi-phase systems such as the four-phase systems described by a 4D
transformation for analysis. However, it is not within the discussion of this book.

2.3.2 Basic Three-Phase Converters, 3D Coordinates and Their Comparisons

The literatures on controlling three-phase three-wire converters are rich; recently
more and more attention is paid to the three-phase four-wire converters [23–27, 30].
In this section, in contrast to the earlier approaches, the comparisons among the
basic three-phase converters of their voltage vectors in 3-dimensional coordinates
are performed, rather than using the switching tables only. These basic converters
are of three-leg, three-leg center-split and four-leg. All the corresponded converters
are illustrated in 2-level equivalent circuits, as shown in Figs. 2.23, 2.24 and 2.25.
In this section, $v_{dc} = v_{dc1} = v_{dc2}$ is assumed for comparisons and none of the
switching losses are being considered.

2.3.2.1 Two-Level Three-Leg Converters

From Sect. 2.3.1, it is found that all the voltage vectors of three-phase converters
can be analyzed by a 3-D transformation basis. According to Fig. 2.23, the
switching functions can be given as $S_j = \{+1, -1\}$, where $j = \{a, b, c\}$. For
example, when $S_a = +1$, S_{a1} is turned on and S_{a2} is turned off, and vice versa.

Fig. 2.23 Two-level three-leg converters

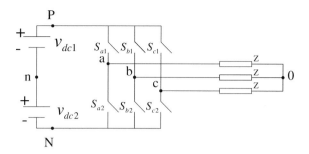

Fig. 2.24 Two-level three-leg center-split converters

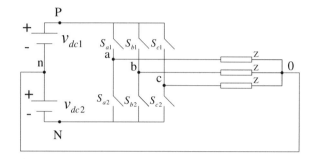

Fig. 2.25 Two-level four-leg converters

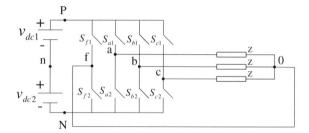

The converter output voltages in a-b-c coordinates can be expressed in (2.53). It is required to transform the voltages from a-b-c into α-β-0 coordinates for comparisons through (2.54).

$$\begin{bmatrix} v_{a0} \\ v_{b0} \\ v_{c0} \end{bmatrix} = \frac{1}{6}(2v_{dc}) \begin{bmatrix} 2 & -1 & -1 \\ -1 & 2 & -1 \\ -1 & -1 & 2 \end{bmatrix} \begin{bmatrix} S_a \\ S_b \\ S_c \end{bmatrix} \tag{2.53}$$

where $S_j = \{+1, -1\}$ and $j = \{a, b, c\}$

$$\begin{bmatrix} v_{\alpha 0} \\ v_{\beta 0} \\ v_{00} \end{bmatrix} = \sqrt{\frac{2}{3}} \begin{bmatrix} 1 & -\frac{1}{2} & -\frac{1}{2} \\ 0 & \frac{\sqrt{3}}{2} & -\frac{\sqrt{3}}{2} \\ \frac{1}{\sqrt{2}} & \frac{1}{\sqrt{2}} & \frac{1}{\sqrt{2}} \end{bmatrix} \begin{bmatrix} v_{a0} \\ v_{b0} \\ v_{c0} \end{bmatrix} \tag{2.54}$$

Furthermore, (2.53) can be inserted into (2.54), which gets (2.55).

$$\begin{aligned} \begin{bmatrix} v_{\alpha 0} \\ v_{\beta 0} \\ v_{00} \end{bmatrix} &= \sqrt{\frac{2}{3}} \frac{(2v_{dc})}{6} \begin{bmatrix} 1 & -\frac{1}{2} & -\frac{1}{2} \\ 0 & \frac{\sqrt{3}}{2} & -\frac{\sqrt{3}}{2} \\ \frac{1}{\sqrt{2}} & \frac{1}{\sqrt{2}} & \frac{1}{\sqrt{2}} \end{bmatrix} \begin{bmatrix} 2 & -1 & -1 \\ -1 & 2 & -1 \\ -1 & -1 & 2 \end{bmatrix} \begin{bmatrix} S_a \\ S_b \\ S_c \end{bmatrix} \\ &= v_{dc} \sqrt{\frac{2}{3}} \begin{bmatrix} 1 & -1/2 & -1/2 \\ 0 & \sqrt{3}/2 & -\sqrt{3}/2 \\ 0 & 0 & 0 \end{bmatrix} \begin{bmatrix} S_a \\ S_b \\ S_c \end{bmatrix} \end{aligned} \tag{2.55}$$

As a result, an instantaneous converter output voltage vector of a three-leg converter can be expressed as (2.56) so that the scalars, K_{α}, K_{β} and K_0, are $\sqrt{2/3} v_{dc} S_{\alpha}$, $1/\sqrt{2} v_{dc} S_{\beta}$ and 0 correspondingly. It should be noticed that v_{dc} is the half value of v_{PN} as shown in Fig. 2.23. There is no zero-component coincidentally, which changes 3-D a-b-c coordinates on 2-D α-β plane.

$$\vec{v}_{three-leg} = v_{dc} \left[\sqrt{\frac{2}{3}} S_{\alpha} \cdot \vec{n}_{\alpha} + \frac{1}{\sqrt{2}} S_{\beta} \cdot \vec{n}_{\beta} \right] \tag{2.56}$$

where $S_{\alpha} = S_a - \frac{1}{2} S_b - \frac{1}{2} S_c$ and $S_{\beta} = S_b - S_c$.

2.3.2.2　Two-Level Three-Leg Center-Split Converters

Comparing Fig. 2.24 with Fig. 2.23, node "O" is connected with node "n". According to the defined switching functions, the voltages are defined as (2.57) in a-b-c coordinates. It forms a two-level three-leg center-split converter shown in Fig. 2.24.

$$\begin{bmatrix} v_{a0} \\ v_{b0} \\ v_{c0} \end{bmatrix} = v_{dc} \begin{bmatrix} 1 & 0 & 0 \\ 0 & 1 & 0 \\ 0 & 0 & 1 \end{bmatrix} \begin{bmatrix} S_a \\ S_b \\ S_c \end{bmatrix} \tag{2.57}$$

where $S_j = \{+1, -1\}$ and $j = \{a, b, c\}$

Furthermore, the converter output voltages in α-β-0 coordinates are expressed in (2.58). An instantaneous converter output voltage vector of a three-leg center-split converter can be represented by (2.59).

$$
\begin{bmatrix} v_{\alpha 0} \\ v_{\beta 0} \\ v_{00} \end{bmatrix} = v_{dc}\sqrt{\frac{2}{3}}
\begin{bmatrix} 1 & -\frac{1}{2} & -\frac{1}{2} \\ 0 & \frac{\sqrt{3}}{2} & -\frac{\sqrt{3}}{2} \\ \frac{1}{\sqrt{2}} & \frac{1}{\sqrt{2}} & \frac{1}{\sqrt{2}} \end{bmatrix}
\begin{bmatrix} 1 & 0 & 0 \\ 0 & 1 & 0 \\ 0 & 0 & 1 \end{bmatrix}
\begin{bmatrix} S_a \\ S_b \\ S_c \end{bmatrix}
$$

$$
= v_{dc}\sqrt{\frac{2}{3}}
\begin{bmatrix} 1 & -1/2 & -1/2 \\ 0 & \sqrt{3}/2 & -\sqrt{3}/2 \\ 1/\sqrt{2} & 1/\sqrt{2} & 1/\sqrt{2} \end{bmatrix}
\begin{bmatrix} S_a \\ S_b \\ S_c \end{bmatrix}
\tag{2.58}
$$

In this case, scalars, K_α, K_β and K_0, are $\sqrt{2/3}v_{dc}S_\alpha$, $1/\sqrt{2}v_{dc}S_\beta$ and $1/\sqrt{3}v_{dc}S_0$ correspondingly.

$$
\vec{v}_{center-split} = v_{dc}\left[\sqrt{\frac{2}{3}}S_\alpha \cdot \vec{n}_\alpha + \frac{1}{\sqrt{2}}S_\beta \cdot \vec{n}_\beta + \frac{1}{\sqrt{3}}S_0 \cdot \vec{n}_0\right]
\tag{2.59}
$$

where $S_\alpha = S_a - \frac{1}{2}S_b - \frac{1}{2}S_c$, $S_\beta = S_b - S_c$ and $S_0 = S_a + S_b + S_c$.

2.3.2.3   Two-Level Four-Leg Converters

There is a growing interest in the four-leg converters for three-phase four-wire applications. Figure 2.25 shows the equivalent circuits of a four-leg converter. The mid-point "f" of an addition leg is connected with node "O". When switching functions are only considered with +1 and −1, there are 2 operation modes for the four-leg converter, as shown in Figs. 2.26 and 2.27.

Fig. 2.26 Mode 1 when switching function of S_f is +1

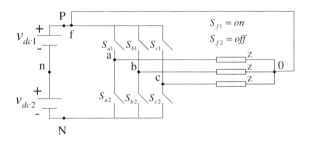

Fig. 2.27 Mode 2 when switching function of S_f is −1

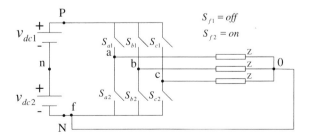

Fig. 2.28 One-leg converter
in Mode 2

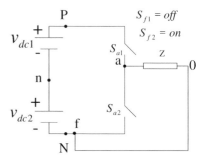

When S_f is +1, the node "O" is connected with node "P". In another mode, node "O" is connected with node "N" if S_f is −1.

For simplification only one leg of them is considered in mode 2 as shown in Fig. 2.28. It has $v_{a0} = v_{dc1} + v_{dc2} = 2v_{dc}$. When $S_{a1} = 1$, $v_{a0} = 2v_{dc}$. In another way, when $S_a = -1$, $v_{a0} = 0$. As a result, the instantaneous voltage can be deduced as (2.60).

$$v_{a0} = (S_a - S_f) \cdot \frac{2v_{dc}}{2} \tag{2.60}$$

As a result, the instantaneous converter output voltages can be defined in a-b-c coordinates as shown in (2.61). The four-leg inverter can produce three independent output voltages, regardless of load. Furthermore, the instantaneous converter output voltages in α-β-0 coordinates from a-b-c coordinates are expressed in (2.62). An instantaneous converter output voltage vector of a four-leg converter can be represented by (2.63). Scalars, K_α, K_β and K_0, are $\sqrt{2/3}v_{dc}S_\alpha$, $1/\sqrt{2}v_{dc}S_\beta$ and $1/\sqrt{3}v_{dc}S_0$ correspondingly.

$$\begin{bmatrix} v_{a0} \\ v_{b0} \\ v_{c0} \end{bmatrix} = v_{dc} \begin{bmatrix} 1 & 0 & 0 & -1 \\ 0 & 1 & 0 & -1 \\ 0 & 0 & 1 & -1 \end{bmatrix} \begin{bmatrix} S_a \\ S_b \\ S_c \\ S_f \end{bmatrix} \tag{2.61}$$

$$\begin{bmatrix} v_{\alpha0} \\ v_{\beta0} \\ v_{00} \end{bmatrix} = \sqrt{\frac{2}{3}}v_{dc} \begin{bmatrix} 1 & -\frac{1}{2} & -\frac{1}{2} \\ 0 & \frac{\sqrt{3}}{2} & -\frac{\sqrt{3}}{2} \\ \frac{1}{\sqrt{2}} & \frac{1}{\sqrt{2}} & \frac{1}{\sqrt{2}} \end{bmatrix} \begin{bmatrix} 1 & 0 & 0 & -1 \\ 0 & 1 & 0 & -1 \\ 0 & 0 & 1 & -1 \end{bmatrix} \begin{bmatrix} S_a \\ S_b \\ S_c \\ S_f \end{bmatrix}$$

$$= \sqrt{\frac{2}{3}}v_{dc} \begin{bmatrix} 1 & -1/2 & -1/2 & 0 \\ 0 & \sqrt{3}/2 & -\sqrt{3}/2 & 0 \\ 1/\sqrt{2} & 1/\sqrt{2} & 1/\sqrt{2} & -3/\sqrt{2} \end{bmatrix} \begin{bmatrix} S_a \\ S_b \\ S_c \\ S_f \end{bmatrix} \tag{2.62}$$

Fig. 2.29 Defined parameters in coordinates

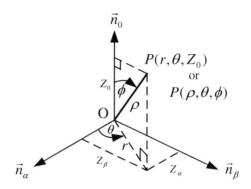

$$\vec{v}_{four-leg} = v_{dc} \left[\sqrt{\frac{2}{3}} S_\alpha \cdot \vec{n}_\alpha + \frac{1}{\sqrt{2}} S_\beta \cdot \vec{n}_\beta + \frac{1}{\sqrt{3}} S_0 \cdot \vec{n}_0 \right] \qquad (2.63)$$

where $S_\alpha = S_a - \frac{1}{2}S_b - \frac{1}{2}S_c$, $S_\beta = S_b - S_c$ and $S_0 = S_a + S_b + S_c - 3S_f$.

2.3.2.4 Three-Phase Converters, Voltage Space Vectors and Their Comparisons

In 3-D aspect, a vector or a switching state can be represented by Rectangular, Cylindrical or Spherical Coordinates. According to parameters defined in Fig. 2.29, the instantaneous converter output voltage vector can be further considered as follows.

$$\vec{v} = v_{dc} \cdot \left(Z_\alpha \vec{n}_\alpha + Z_\beta \vec{n}_\beta + Z_0 \vec{n}_0 \right) \qquad (2.64)$$

where $Z_\alpha = \sqrt{\frac{2}{3}}S_\alpha$, $Z_\beta = \frac{1}{\sqrt{2}}S_\beta$ and $Z_0 = \frac{1}{\sqrt{3}}S_0$.

The parameters defined in Fig. 2.29 can be expressed as:

$$\rho = \sqrt{Z_\alpha^2 + Z_\beta^2 + Z_0^2} \qquad (2.65)$$

$$r = \sqrt{Z_\alpha^2 + Z_\beta^2} \qquad (2.66)$$

$$\theta = \tan^{-1}\left(\frac{Z_\beta}{Z_\alpha}\right) \qquad (2.67)$$

$$\phi = \cos^{-1}\left(\frac{Z_0}{\rho}\right) = \sin^{-1}\left(\frac{r}{\rho}\right) \qquad (2.68)$$

Table 2.2 Two-level three-leg converter voltage vector parameters

	S_a	S_b	S_c	S_α	S_β	S_0	r	θ	Z_0	ρ	ϕ
$\vec{V}_1$	1	−1	−1	2	0	0	1.633	0°	0	1.633	90°
$\vec{V}_2$	1	1	−1	1	2	0	1.633	60°	0	1.633	90°
$\vec{V}_3$	−1	1	−1	−1	2	0	1.633	120°	0	1.633	90°
$\vec{V}_4$	−1	1	1	−2	0	0	1.633	180°	0	1.633	90°
$\vec{V}_5$	−1	−1	1	−1	−2	0	1.633	240°	0	1.633	90°
$\vec{V}_6$	1	−1	1	1	−2	0	1.633	300°	0	1.633	90°
$\vec{V}_{0p}$	1	1	1	0	0	0	0	*	0	0	90°
$\vec{V}_{0n}$	−1	−1	−1	0	0	0	0	*	0	0	90°

However, $v_{dc} = 1$ is assumed for simplifying the comparison among three-phase converters as follows.

A. A Two-Level Three-Leg Converter and Its Voltage Space Vector

In a two-level three-leg converter, there are $2^3 = 8$ available voltage vectors but with only 7 achievable switching states. Coincidentally, there is no zero sequence component of voltage space vectors in α-β-0 coordinates. The available vectors are projected from 3D a-b-c coordinates on a 2D α-β plane. Table 2.2 summarizes parameters of the 2-level voltage vectors for a two-level three-leg converter. The mark, *, means "undefined" in following Tables. Figure 2.30 shows the allocation of available output voltage vectors of a two-level three-leg converter.

In conventional 2-D Pulse Width Modulation techniques, voltage space vector allocation can be illustrated on a α-β plane, which is shown in Fig. 2.30. There are totally 8 available voltage vectors with 6 directional vectors and 2 zero vectors. These 2 zero vectors are utilized in optimizing the switching sequence in order to reduce the switching loss.

Fig. 2.30 Allocation of available output voltage vectors of a two-level three-leg converter

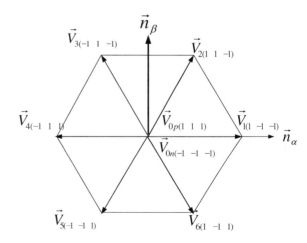

Table 2.3 Two-level three-leg center-split converter voltage vector parameters

	S_a	S_b	S_c	S_α	S_β	S_0	r	θ	Z_0	ρ	ϕ
$\vec{V}_1$	1	−1	−1	2	0	−1	1.633	0°	−0.577	$\sqrt{3}$	109.46°
$\vec{V}_2$	1	1	−1	1	2	1	1.633	60°	0.577	$\sqrt{3}$	70.54°
$\vec{V}_3$	−1	1	−1	−1	2	−1	1.633	120°	−0.577	$\sqrt{3}$	109.46°
$\vec{V}_4$	−1	1	1	−2	0	1	1.633	180°	0.577	$\sqrt{3}$	70.54°
$\vec{V}_5$	−1	−1	1	−1	−2	−1	1.633	240°	−0.577	$\sqrt{3}$	109.46°
$\vec{V}_6$	1	−1	1	1	−2	1	1.633	300°	0.577	$\sqrt{3}$	70.54°
$\vec{V}_{0p}$	1	1	1	0	0	3	0	*	$\sqrt{3}$	$\sqrt{3}$	0°
$\vec{V}_{0n}$	−1	−1	−1	0	0	−3	0	*	−$\sqrt{3}$	$\sqrt{3}$	180°

B. A Two-Level Three-Leg Center-Split Converter and Its Voltage Space Vector

In a two-level three-leg center-split converter, there are $2^3 = 8$ available voltage vectors with 8 achievable switching states. The available vectors are projected from 3D *a-b-c* coordinates into 3D *α-β-0* coordinates. Table 2.3 summarizes the voltage vector parameters for a two-level three-leg center-split converter. Figure 2.31 shows the allocation of available output voltage vectors of a two-level three-leg center-split converter.

Comparing Figs. 2.30 and 2.31, it shows that the number of available voltage vectors is equal to the number of available states in a two-level three-leg center-split converter, which is not the case in a two-level three-leg converter. Those two zero

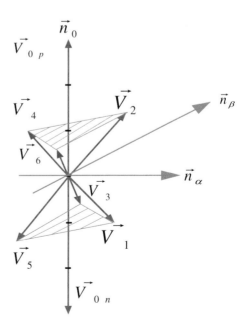

Fig. 2.31 Allocation of available output voltage vectors of a two-level three-leg center-split converter

vectors can be employed for a neutral line current compensation as well as optimizing the switching sequence to reduce switching loss. All the vectors of the three-leg center-split converter have zero components. It is noticed that the amplitude, r, of $\{\vec{V}_1, \vec{V}_2, \vec{V}_3, \vec{V}_4, \vec{V}_5, \vec{V}_6\}$ vectors of the three-leg center-split converter is the same as that of the original three-leg converter, which is the projection of $\{\vec{V}_1, \vec{V}_2, \vec{V}_3, \vec{V}_4, \vec{V}_5, \vec{V}_6\}$ vectors on the α-β plane.

C. A Two-Level Four-Leg Converter and Its Voltage Space Vector

In a two-level four-leg converter, there are $2^4 = 16$ available vectors with 15 achievable switching states. The available voltage vectors are projected from 3D a-b-c coordinates into 3D α-β-0 coordinates. Table 2.4 summarizes the voltage vector parameters for a two-level four-leg converter. Figure 2.32 shows the output voltage allocation of available vectors of a two-level four-leg converter. The amplitude, r, of $\{\vec{V}_{1p}, \vec{V}_{2p}, \vec{V}_{3p}, \vec{V}_{4p}, \vec{V}_{5p}, \vec{V}_{6p}, \vec{V}_{1n}, \vec{V}_{2n}, \vec{V}_{3n}, \vec{V}_{4n}, \vec{V}_{5n}, \vec{V}_{6n}\}$ vectors of the four-leg converter is the same as that of the original three-leg and the three-leg center-split converters, which are the projections of those vectors on the α-β plane.

D. Comparisons and Discussions

Table 2.5 summarizes the above analysis results. In previous studies, under the center-split capacitor approach, the three-phase converter essentially becomes three single-phase half-bridge converters. Thus, it suffers a lower utilization of dc link

Table 2.4 Two-level four-leg converter voltage vector parameters

	S_f	S_a	S_b	S_c	S_α	S_β	S_0	r	θ	Z_0	ρ	ϕ
$\vec{V}_{1p}$	-1	1	-1	-1	2	0	2	1.633	0°	1.155	2	54.73°
$\vec{V}_{2p}$	-1	1	1	-1	1	2	4	1.633	60°	2.309	2.828	35.27°
$\vec{V}_{3p}$	-1	-1	1	-1	-1	2	2	1.633	120°	1.155	2	54.73°
$\vec{V}_{4p}$	-1	-1	1	1	-2	0	4	1.633	180°	2.309	2.828	35.27°
$\vec{V}_{5p}$	-1	-1	-1	1	-1	-2	2	1.633	240°	1.155	2	54.73°
$\vec{V}_{6p}$	-1	1	-1	1	1	-2	4	1.633	300°	2.309	2.828	35.27°
$\vec{V}_{0p}$	-1	1	1	1	0	0	6	0	*	3.464	3.464	0°
$\vec{V}_{00n}$	-1	-1	-1	-1	0	0	0	0	*	0	0	90°
$\vec{V}_{1n}$	1	1	-1	-1	2	0	-4	1.633	0°	-2.309	2.828	144.73°
$\vec{V}_{2n}$	1	1	1	-1	1	2	-2	1.633	60°	-1.155	2	125.27°
$\vec{V}_{3n}$	1	-1	1	-1	-1	2	-4	1.633	120°	-2.309	2.828	144.73°
$\vec{V}_{4n}$	1	-1	1	1	-2	0	-2	1.633	180°	-1.155	2	125.27°
$\vec{V}_{5n}$	1	-1	-1	1	-1	-2	-4	1.633	240°	-2.309	2.828	144.73°
$\vec{V}_{6n}$	1	1	-1	1	1	-2	-2	1.633	300°	-1.155	2	125.27°
$\vec{V}_{0n}$	1	-1	-1	-1	0	0	-6	0	*	-3.464	-3.464	180°
$\vec{V}_{00p}$	1	1	1	1	0	0	0	0	*	0	0	90°

Fig. 2.32 Allocation of available output voltage vectors of a two-level four-leg converter

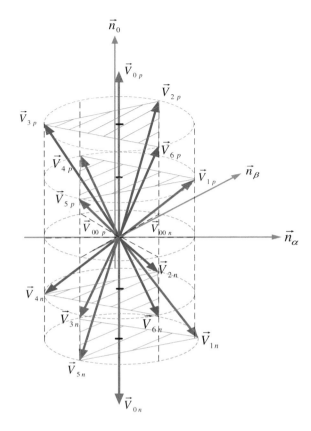

Table 2.5 Comparisons among three basic three-phase converters

	Maximum phase voltage	Voltage capacity on α-β plane	Voltage capacity on zero axis
Three-leg converters	$\frac{4}{3}v_{dc}$	Same	0
Three-leg center-split converters	v_{dc}	Same	Middle
Four-leg converters	$2v_{dc}$	Same	Highest

voltage as compared to the four-leg. Based on the maximum phase voltage that could be performed by the three-phase converters, reaching the above conclusion was straightforward. However, according to Tables 2.2, 2.3 and 2.4, all of the three-phase converters have the same amplitude, r, of the projected non-zero vectors on the α-β plane. It means that when the converters have the same dc linked voltage, they have the same α-β voltage capacity for converter's operation, and the four-leg converter has higher zero-current injection ability as it has a higher Z_0 than that of the three-leg center-split. As a result, a three-leg center split and four-leg inverters have the same voltage capacity on the α-β plane when they have the same dc linked voltage.

Table 2.6 Simulation results

$v_{dc} = v_{dc1} = v_{dc2} = 100$ V $v_{PN} = 200$ V	Phase voltage (V)	$r \cdot v_{dc}$ (V)	$Z_0 \cdot v_{dc}$ (V)
Three-leg converters $[S_a\ S_b\ S_c] = [1]$	133.3	163.3	1.641×10^{-14}
Three-leg center-split converters $[S_a\ S_b\ S_c] = [1]$	99.99	163.3	-57.73
Four-leg converters $[S_a\ S_b\ S_c\ S_f] = [1]$	200	163.3	115.4

A simulation test of the three-phase three-leg, the three-leg center-split and the four-leg inverters is performed when $v_{dc} = v_{dc1} = v_{dc2} = 100$ V and $v_{PN} = v_{dc1} + v_{dc2} = 200$ V with balanced loads are connected. The switching functions are given by Table 2.6. It is noticed that the total dc linked voltage is 200 V. Table 2.6 summarizes the simulation results, where r and Z_0 are defined in Tables 2.2, 2.3 and 2.4 respectively. All the simulated values followed the above theoretical results. It shows the validity of the above theoretical conclusions. First, phase voltage is the highest in the four-leg inverter and the three-leg center-split converter has the smallest phase voltage. Second, all the inverters have the same voltage capacity on the α-β plane when they have the same dc linked voltage. Third, the four-leg inverter has the highest zero injection capacity. Fourth, only one transformation can be employed to analyze all three-phase converters.

In left columns of Tables 2.2, 2.3 and 2.4, the switching functions in *a-b-c* and *α-β-0* coordinates are identical to those in the other researches [23–30]. Experimental results were performed in [25, 32] to show the validity of switching function tables. However, without using a 3D voltage model, such as (2.56), (2.59) and (2.63), to describe the output vectors of three-phase converters, it was hard to compare the relationship among the converters. Based on the mathematical and theoretical deduction with only one 3D transformation basis, the converters can be comparable directly.

2.3.3 DC Voltage Unbalance, Variation, Switching Functions and 3D Coordinates

In the three-phase converters, the unbalance in dc-link neutral point voltages, for example, in a three-phase three-leg center-split converter, may affect PWM performance and it may cause a significant distortion in generating a current waveform. In this section, $v_{dc1} \neq v_{dc2}$ is assumed in order to consider the influence of the unbalance dc upper and lower-leg voltage variations on 3D voltage space vectors of the three-phase converters. Switching functions are considered due to the unbalance between dc upper and lower capacitor voltages. V_{dc} is defined as (2.69). The upper-leg and lower-leg switching functions in a, b or c phases can be defined as

(2.70) accordingly, where S_j^N is defined as this new proposed switching function due to dc voltage unbalance and j = a, b or c.

$$V_{dc} = \frac{v_{dc1} + v_{dc2}}{2} \qquad (2.69)$$

$$S_j^N = \begin{cases} \frac{v_{dc1}}{V_{dc}}, & \textit{upper leg} \\ -\frac{v_{dc2}}{V_{dc}}, & \textit{lower leg} \end{cases} \qquad (2.70)$$

Conventionally, the switching function can be 1 or −1, but this proposed switching function can either be smaller or larger than 1 due to the dc voltage deviation. When $v_{dc1} = v_{dc2} = V_{dc}$, $S_j^N = S_j$. However, when $v_{dc1} \neq v_{dc2} \neq V_{dc}$, $S_j^N \neq \pm 1$. Furthermore, S_j^N can be expressed as (2.71). The proposed switching function is found to be affected by the dc voltage unbalance.

$$S_j^N = \begin{cases} \frac{v_{dc1}}{V_{dc}} = \frac{2v_{dc1}}{v_{dc1}+v_{dc2}} = \frac{(v_{dc1}+v_{dc2})+(v_{dc1}-v_{dc2})}{v_{dc1}+v_{dc2}} = 1 + \Delta S, & \textit{upper leg} \\ -\frac{v_{dc2}}{V_{dc}} = \frac{-2v_{dc2}}{v_{dc1}+v_{dc2}} = \frac{-(v_{dc1}+v_{dc2})+(v_{dc1}-v_{dc2})}{v_{dc1}+v_{dc2}} = -1 + \Delta S, & \textit{lower leg} \end{cases} \qquad (2.71)$$

where $\Delta S = \frac{v_{dc1}-v_{dc2}}{2V_{dc}}$.

The relationship between the new switching function and the original one can be expressed as:

$$S_j^N = S_j + \Delta S \qquad (2.72)$$

Therefore, the output voltage vector of a three-leg inverter can be recalculated by S_a^N, S_b^N and S_c^N due to the dc voltage unbalance. And, V_{dc} can either be larger or smaller due to the dc voltage variation.

2.3.3.1 A Three-Leg Converter

The new instantaneous voltage vector $\vec{v}^N$ can be expressed as (2.73) due to the dc upper-leg and lower-leg voltage unbalance.

$$\vec{v}_{three-leg}^N = V_{dc} \left[\sqrt{\frac{2}{3}} S_\alpha^N \cdot \vec{n}_\alpha + \frac{1}{\sqrt{2}} S_\beta^N \cdot \vec{n}_\beta \right] \qquad (2.73)$$

where $S_\alpha^N = S_a^N - \frac{1}{2}S_b^N - \frac{1}{2}S_c^N$

$$S_\beta^N = S_b^N - S_c^N$$

Substituting (2.72) into (2.73), no influence is found on the space vector voltage of a three-leg converter, that is $\vec{v}_{three-leg} = \vec{v}^N_{three-leg}$. However, the instantaneous voltage bounded area can either be larger or smaller due to the dc voltage variation, V_{dc}.

2.3.3.2 A Three-Leg Center-Split Converter

The instantaneous voltage vector $\vec{v}^N$ of a three-leg center-split converter in 3D can be expressed as (2.74).

$$\vec{v}^N_{center-split} = V_{dc}\left[\sqrt{\frac{2}{3}}S^N_\alpha \cdot \vec{n}_\alpha + \frac{1}{\sqrt{2}}S^N_\beta \cdot \vec{n}_\beta + \frac{1}{\sqrt{3}}S^N_0 \cdot \vec{n}_0\right] \tag{2.74}$$

where $S^N_\alpha = S^N_a - \frac{1}{2}S^N_b - \frac{1}{2}S^N_c$

$$S^N_\beta = S^N_b - S^N_c$$

$$S^N_0 = S^N_a + S^N_b + S^N_c$$

Substituting (2.72) into (2.74), only the zero switching function is found to be affected by the dc voltage unbalance.

$$S^N_\alpha = S_\alpha \tag{2.75}$$

$$S^N_\beta = S_\beta \tag{2.76}$$

$$S^N_0 = S_0 + 3\Delta S \tag{2.77}$$

The instantaneous voltage vector $\vec{v}^N_{center-split}$ can be expressed as (2.68).

$$\vec{v}^N_{center-split} = \vec{v}_{cemter-split} + \vec{v}^N_0 \tag{2.78}$$

where $\vec{v}^N_0 = \sqrt{3}V_{dc}\Delta S \cdot \vec{n}_0$.

According to (2.78), the voltage vectors may vary upward or downward along $\vec{n}_0$ axis due to the unbalance of dc voltage variation as shown in Fig. 2.33. However, the instantaneous voltage bounded volume can either be larger or smaller due to the dc voltage variation, V_{dc}.

Fig. 2.33 2-Level 3D vector allocation due to dc voltage unbalance

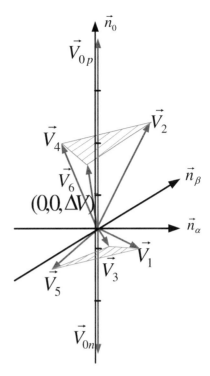

2.3.3.3 A Four-Leg Converter

The instantaneous voltage vector $\vec{v}^N$ of a four-leg converter in 3D can be expressed as (2.79).

$$\vec{v}^N_{four-leg} = V_{dc} \left[\sqrt{\frac{2}{3}} S^N_\alpha \cdot \vec{n}_\alpha + \frac{1}{\sqrt{2}} S^N_\beta \cdot \vec{n}_\beta + \frac{1}{\sqrt{3}} S^N_0 \cdot \vec{n}_0 \right] \tag{2.79}$$

where $S^N_\alpha = S^N_a - \frac{1}{2} S^N_b - \frac{1}{2} S^N_c$

$$S^N_\beta = S^N_b - S^N_c$$

$$S^N_0 = S^N_a + S^N_b + S^N_c - 3 S^N_f$$

Substituting (2.72) into (2.79), it is found that there is no influence on the space vector voltages of a four-leg converter due to dc voltage unbalance, which is $\vec{v}_{four-leg} = \vec{v}^N_{four-leg}$. However, the instantaneous voltage bounded volume can either be larger or smaller due to the dc voltage variation, V_{dc}.

From the above calculations, only the space voltage vectors of the three-leg center-split converter are affected due to the dc voltage unbalance. Reversely, the dc unbalance voltages can be controlled by shifting upward and downward of the reference voltage vector accordingly. Details and experimental results for a three-leg center-split converter can be found in [32]. However, the instantaneous voltage bounded volume can either be larger or smaller due to the dc voltage variation, V_{dc}.

2.4 Summary

Three fundamental three-phase converter circuit topologies are investigated and compared under only one coordinate transformation and a 3D model proposed in this chapter. Several conclusions are drawn:

(1) Phase voltage is the highest in a four-leg inverter and a three-leg center-split converter has the smallest phase voltage.
(2) All the inverters have the same voltage capacity on α-β plane when they have the same dc-linked voltage.
(3) A four-leg inverter has the highest zero injection capacity.
(4) Switching functions can be defined either smaller or larger than 1 due to the dc voltage unbalance.
(5) Space voltage vectors of a three-leg center-split converter are affected due to the dc voltage unbalance by shifting upward and downward of the reference voltage vector along zero-axis accordingly.
(6) Space voltage vectors bounded volume can either be larger or smaller due to dc voltage variation.

References

1. S.T. Senini, P.J. Wolfs, Systematic identification and review of hybrid active filter topologies, in *Proceedings IEEE 33rd Annual Power Electronics Specialists Conference, PESC 02*, vol. 1, pp. 394–399 (2002)
2. H. Fujita, H. Akagi, A practical approach to harmonic compensation in power systems—series connection of passive and active filters. IEEE Trans. Indus. Appl. **27**, 1020–1025 (1991)
3. F.Z. Peng, H. Akagi, A. Nabae, A new approach to harmonic compensation in power systems – a combined system of shunt passive and series active filters. IEEE Trans. Indus. Appl. **26**, 983–990 (1990)
4. L. Chen, A.V. Jouanne, A comparison and assessment of hybrid filter topologies and control algorithms, in *Proceedings IEEE 32nd Annual Power Electronics Specialists Conference, PESC. 01*, vol. 2, pp. 565–570 (2001)
5. D.A. Gonzalez, J.C. McCall, Design of filters to reduce harmonic distortion in industrial power system. IEEE Trans. Indus. Appl. **23**(3), 504–511 (1987)

6. L. Morán, J. Dixon, R. Wallace, A three-phase active power filter operating with fixed switching frequency for reactive power and current harmonic compensation. IEEE Trans. Indus. Electron. **42**, 402–408 (1996)
7. C.-S. Lam, M.-C. Wong, A novel b-shaped L-type transformerless hybrid active power filter in three-phase four-wire systems, in *The 38th North American Power Symposium*, SIU Carbondale, Illinois, Sept. 17–19, 2006, pp. 235–241
8. D. Rivas, L. Moran, J.W. Dixon, J.R. Espinoza, Improving passive filter compensation performance with active techniques. IEEE Trans. Indus. Electron. **50**(1), 161–170 (2003)
9. L. Gyuyi, E.C. Strycula, Active ac Power Filter, in *Proceeding of 1976 IEEE/IAS Annual Meeting*, Chicago, Illinois, Oct. 11–14, 1976
10. B. Singh, K. Al-Haddad, A. Chandra, A review of active filters for power quality improvement. IEEE Trans. Indus. Electron. **46**(5), 960–972 (1999)
11. H. Akagi, New trends in active filters for power conditioning. IEEE Trans. Indus. Appl. **32**(6), 1312–1322 (1996)
12. J.-S. Lai, T.S. Key, Effectiveness of harmonic mitigation equipment for commercial office buildings. IEEE Trans. Indus. Appl. **33**(4), 1104–1110 (1997)
13. E.H. Watanabe, H. Akagi, M. Aredes, Instantaneous p-q power theory for compensating nonsinusoidal systems, in *International School on Nonsinusoidal Currents and Compensation (ISNCC)*, Lagow, Poland, pp. 1–10, June 10–13, 2008
14. H. Akagi, Y. Kanazawa, A. Nabae, Instantaneous reactive power compensators comprising switching devices without energy storage components. IEEE Trans. Indus. Appl. **20**, 625–630 (1984)
15. A. Nabae, T. Tanaka, A new definition of instantaneous active-reactive current and power based on instantaneous space vectors on polar coordinates in three-phase circuits, in *IEEE/PES Winter Meeting*, Baltimore, Maryland, Jan 21–25, 1996, pp. 1238–1243
16. J.L. Willems, A new interpretation of the Akagi-Nabae power components of nonsinusoidal three-phase situation. IEEE Trans. Instrum. Meas. **41**(4), 523–527 (1992)
17. A. Nabae, et al., Reactive power compensation based on neutral line current separating and combining method. Trans. IEE Jpn. 114-D(7/8), 800–801 (1994) (in Japanese)
18. F.Z. Peng, J.S. Lai, Generalized instantaneous reactive power theory for three-phase power systems. IEEE Trans. Instrum. Meas. **45**(1), 293–297 (1996)
19. F.Z. Peng, G.W. Ott Jr, D.J. Adams, Harmonic and reactive power compensation based on the generalized instantaneous reactive power theory for three-phase four-wire systems. IEEE Trans. Power Electron. **13**(6), 1174–1181 (1998)
20. S.-K. Chen, G. W. Chang, A new instantaneous power theory-based three-phase active power filter, in *IEEE Power Engineering Society Winter Meeting, Singapore*, vol. 4, Jan. 23–27, 2000, pp. 2687–2692
21. H. Akagi, Active filters and energy storage systems operated under non-periodic conditions, in *IEEE Power Engineering Society Summer Meeting*, vol. 2, Washington, USA, July 16–20, 2000, pp. 965–970
22. M. Aredes, J. Hafner, K. Heumann, Three-phase four-wire shunt active filter control strategies. IEEE Trans. Power Electron. **12**(2), 311–318 (1997)
23. P. Verdelho, G.D. Marques, Four-wire current regulator PWM voltage converter. IEEE Trans. Indus. Electron. **45**, 761–770 (1998)
24. A. Dastfan, V.J. Gosbell, D. Platt, Control of a new active power filter using 3-D vector control. IEEE Trans. Power Electron. **15**(1), 5–12 (2000)
25. R. Zhang, V.H. Prasad, D. Boroyevich, F.C. Lee, Three-dimensional space vector modulation for four-leg voltage-source converters. IEEE Trans. Power Electron. **17**(3), 314–326 (2002)
26. S.M.A. Marian, P. Kazmierkowski, PWM voltage and current control of four-leg VSI, in *Proceedings of IEEE International Symposium on Industry Electronics*, vol. 1, pp. 196–201, July 1998
27. M.J. Ryan, W. Doncker, R.D. Lorenz, Decoupled control of a four-leg inverter via a new 4×4 transformation matrix. IEEE Trans. Power Electron. **16**(5), 694–701 (2001)

28. M.C. Wong, Z.-Y. Zhao, Y.-D. Han, L.-B. Zhao, Three-dimensional pulse-width modulation technique in three-level power inverters for three-phase four-wired system. IEEE Trans Power Electron **16**(3), 418–427 (2001)
29. M.C. Wong, J. Tang, Y.-D. Han, Cylindrical coordinate control of three-dimensional PWM technique in three-phase four-wired trilevel inverter. IEEE Trans. Power Electron. **18**(3), 208–219 (2003)
30. M.J. Ryan, R.D. Lorenz, R.W. Doncker, Modeling of multileg sine-wave inverters: A geometric approach. IEEE Trans. Indus. Electron. **46**(6), 1183–1191 (1999)
31. H. Anton, *Elementary Linear Algebra* (Wiley, 1997)
32. N.Y. Dai, M.C. Wong, Y.D. Han, Three-dimensional space vector modulation with dc voltage variation control in three-leg center-split power quality compensators, in *IEE Proceedings— Electric Power Applications*, vol. 151, no. 2, pp. 198–204, March 2004

Chapter 3
Active Power Filters

Abstract In this chapter, the active power filters are discussed. In Sect. 3.1, the development of active power filters are reviewed; active power filters with different circuit configurations are given and compared. The three-phase four-wire active power filter using a two level four-leg voltage source inverter (VSI) is discussed in Sect. 3.2. Three-phase four-wire active power filters using two-level and three-level three-leg center-split VSI are discussed in Sect. 3.3. The modeling, voltage control signal calculation, space vector analyses and pulse width modulation (PWM) methods for each configuration are introduced and addressed in these two Sections. Furthermore, dc link voltage variation control of the center-split VSI is discussed in Sect. 3.3. In Sect. 3.4, the operational principal of three-phase four-wire multi-level VSIs as active power filter is briefly introduced. In Sect. 3.5, the generalized PWM for multi-level three-leg center-split and four-leg VSIs are introduced. Finally, the design and implementation of an active power filter is introduced in Sect. 3.6.

3.1 Development of Active Power Filters

The proliferation of a variety of nonlinear loads in utility power system can cause a large amount of harmonics injected into power system. For a long time, L-C tuned filters have been used for suppressing harmonics in ac power system [1]. These filters, however, have inherent drawback, e.g. tuned frequencies are fixed and a harmonic amplifying phenomenon may occur due to parallel resonance between the line inductance and filter capacitance. To solve those problems, more attention has been paid to active power filter (APF) and researches on them are prospering [2, 3].

The active power filter (APF) is one of the shunt connected power quality compensators. The concept, "Active ac Power Filter", was first developed by Gyugyi in 1970s [4]. The APF can overcome the disadvantages inherent in passive power filters (PPF). Since then, active power filters have been becoming the most popular shunt power quality compensator, which can compensate harmonics and reactive current simultaneously. In addition, multiple load operation with one active filter is technically and economically feasible.

© Springer Science+Business Media Singapore 2016
M.-C. Wong et al., *Parallel Power Electronics Filters
in Three-Phase Four-Wire Systems*, DOI 10.1007/978-981-10-1530-4_3

A large number of publications have explored active filters for three-phase three-wire systems with different configurations. With three wires on the ac side and two wires on the dc side, APF works as a current source. By using a static converter, APF generates the compensation currents which are equal in magnitude but out of phase by 180° to the harmonics included in load currents, and are injected into the ac lines. Therefore, the converter is required to have high current control capability and high operating efficiency.

Both voltage source and current source converters can be used in an APF [5]. A current source APF is shown in Fig. 3.1 and the size of required dc-link inductor is large, which may cause electro-magnetic interference (EMI) to nearby electric equipment. A voltage source APF is given in Fig. 3.2, which is superior in efficiency because the loss in a dc capacitor is smaller than that in a dc reactor needed with a current source APF. Moreover, among the existing applications of APFs, ninety percent of them are voltage source APFs.

According to the connection modes, there are four types of APF: Shunt APF, Series APF, Hybrid APF and Series-Shunt APF. However, the series APF connected with the ac lines through three single-phase transformers is equivalent to a controlled voltage source as shown in Fig. 3.3. This APF mainly eliminates the harmonic voltages for the system.

Compared with the shunt APF, the losses, protection, and operation etc. of series APF are more complicated. Now, the single series APF is seldom used. It is always combined with L-C filter that is called hybrid APF as shown in Fig. 3.4. Because the L-C filter can eliminate most of the dominant low order harmonics, the capacity

Fig. 3.1 Current source APF

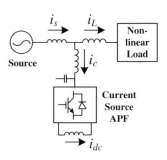

Fig. 3.2 Voltage source APF

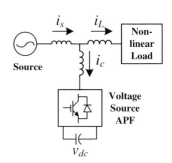

Fig. 3.3 Series APF

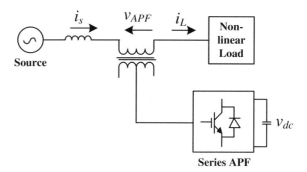

Fig. 3.4 Hybrid APF

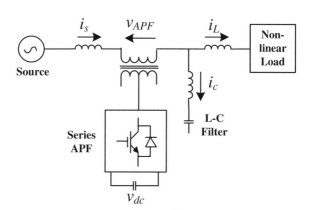

of APF can be very small so thus reducing its volume and cost. Furthermore, because of diminishing both harmonic voltages and harmonic currents simultaneously, it is widely used in the industry applications. However, its main drawback is that it can only operate for certain stable load. When the load is changing rapidly, its effectiveness of compensation can be low.

The last type is series-shunt APF as shown in Fig. 3.5. It is called an unified power quality conditioner (UPQC) and drawn the attention of many researchers recently.

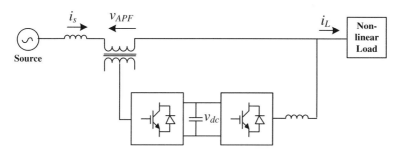

Fig. 3.5 Series-shunt APF

It is composed of series APF and shunt APF together so that it contains functions of both APFs. The series unit can handle the harmonic voltages, voltage fluctuation and flick, and can improve the steadiness of power system, etc. The shunt unit can handle harmonic, unbalance and reactive currents. The series-shunt APF can work in both single-phase and three-phase systems. Until now, it is regarded as the ideal structure of APF. However, the drawback of it is its complicated control and high initial cost.

Figure 3.6 shows a typical APF with a three-phase voltage source inverter (VSI), which is applied in three-phase three-wire systems. It is in parallel connection with ac lines and is equivalent to a controlled current source. It can generate the same value of load reactive and harmonic currents but in opposite direction to compensate those reactive and harmonic currents. The techniques of shunt APF have been widely developed. Therefore, most of the APF applications are based on shunt APF.

Due to the development of "Custom Power" concept, a three-phase four-wire system will play a very important role in the distribution site. In the three-phase four-wire system, there exists neutral current, harmonics, reactive power, and unbalance problems, so the conventional APF designed for three-phase three-wire system cannot compensate all the problems. In 1997, four-leg and three-leg center-split voltage source inverters (VSIs) were proposed to compensate the reactive, unbalance, harmonic and neutral currents for the three-phase four-wire power systems [6]. The three-phase four-wire systems employ mainly four-leg and three-leg center-split VSI for power quality compensation. As a result, only these 2

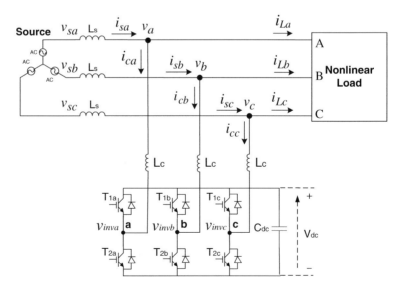

Fig. 3.6 APF uses a three-phase voltage source inverter

Fig. 3.7 A four-leg VSI

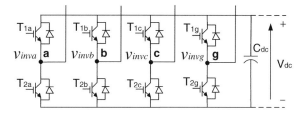

Fig. 3.8 A three-leg center-split VSI

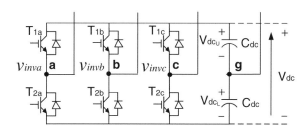

topologies are discussed in this book. The former structure uses a fourth leg to control the neutral current as shown in Fig. 3.7. The latter one is known as the center-split structure, in which the original d.c. capacitor is split into 2 to provide the neutral current compensation as given in Fig. 3.8.

Comparisons were carried out between three-leg centre-split VSIs and four-leg VSIs [7–9]. The two-level four-leg inverter is preferable in low voltage applications. However, multi-level three-leg centre-split VSIs are more preferable. It is because only a neutral wire is connected into multi-level neutral point clamped inverter which can turn the inverter into three-phase four-wire applications without adding extra-switching components. In the following sections, the modeling and control of the three-phase four-wire APFs are provided. A general block diagram of the control system of the APF is shown in the Fig. 3.9. Since the compensating current detection has been discussed in the Chap. 2, the PWM for three-phase four-wire converters are mainly focused in the following sections.

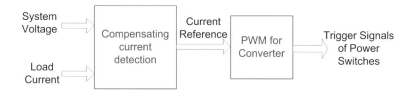

Fig. 3.9 Block diagram of the APF control system

3.2   A Two-Level Four-Leg VSI as a Three-Phase Four-Wire Active Power Filter

A two-level four-leg VSI based APF shown in Fig. 3.10 is an extension of a conventional two-level three-leg VSI based APF, which is added with an extra-leg to the three-leg one in order to provide a neutral current compensation capability. Due to its additional leg, the initial cost of this inverter is higher than that of the three-leg center-split one. However, four-leg VSI is not required to control the dc-voltage unbalance between the upper and lower capacitor voltages of the centre-split structure. The control algorithm of the three-leg centre-split is more complicated than four-leg one. Based on the results from Sect. 2.3.2, although the phase inverter voltage of the four-leg one is higher comparing to that of the three-leg inverters, its effective voltage vector on $\alpha\beta$ plane is the same as the three-leg inverters, which has been explained in Chap. 2. In low power applications, a four-leg inverter is preferable. However in medium or higher power applications, four-leg inverter is not a good choice in multi-level structure development and its related detailed explanation is given in the following sections. As a result, only two-level four-leg inverter is discussed and multi-level four-leg inverter is not included in this book.

3.2.1   Modeling of Two-Level Four-Leg Active Power Filters

A three-phase four-wire active power filter is shown in Fig. 3.10, in which a two-level four-leg VSI is used. To inject the required compensating current, the

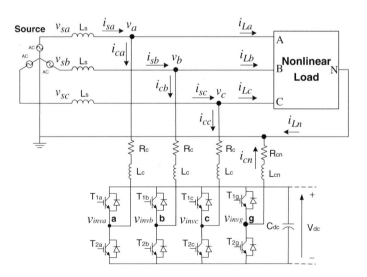

Fig. 3.10 A three-phase four-wire APF uses two-level four-leg VSI

inverter needs to detect the reactive, harmonic and neutral currents of the loading first. The compensating current detection algorithms have been discussed in the Chap. 2. The four-leg VSI is controlled by a pulse width modulation (PWM) to inject the compensating currents into the point of common coupling (PCC). As a result, the reactive current, harmonic and neutral currents can be compensated by the APF. The modeling of the APF with a two-level four-leg VSI is presented in this section.

According to Fig. 3.10, the internal resistance of coupling inductor L_c and L_{cn} are assumed to be very small ($R_c, R_{cn} \approx 0$), which can be neglected, the relation among the compensating currents i_{cx} ($x = $ a, b, c, n), the terminal voltages of VSI v_{invx} ($x = $ a, b, c, g), the terminal voltages at PCC and coupling inductors L_c and L_{cn} can be expressed as:

$$
\begin{cases}
v_{inva} = v_a - L_c \frac{di_{ca}}{dt} \\
v_{invb} = v_b - L_c \frac{di_{cb}}{dt} \\
v_{invc} = v_c - L_c \frac{di_{cc}}{dt} \\
v_{invg} = v_n + L_{cn} \frac{di_{cn}}{dt}
\end{cases}
\tag{3.1}
$$

The switching function of each leg of the VSI is defined in (3.2). As a result, the output voltage of the VSI is given in (3.3).

$$
S_x = \begin{cases} 1 \\ -1 \end{cases} \quad x = a, b, c, g
\tag{3.2}
$$

In Sect. 2.3, referring to Fig. 2.25 and (2.60), one should be noted that $2v_{dc}$ in (2.60) is the total dc linked voltage between P and N in Fig. 2.25 as $v_{dc} = v_{dc1} = v_{dc2}$ was defined in Sect. 2.3. However, in this section, based on Fig. 3.10, v_{dc} is defined as the total dc linked voltage. As a result, (3.3) can be obtained. Chapter 2 was focused on the effective voltage comparison so that $v_{dc} = v_{dc1} = v_{dc2}$ was defined and total dc linked voltage was $2v_{dc}$ for simpler comparison. However, Chap. 3 is focused on control so that the total dc linked voltage is v_{dc}. Comparing instantaneous voltage vectors in Chaps. 2 and 3, such as (2.56), (2.59), (2.63), (3.11), (3.13) and (3.49), one can find that $2v_{dc}$ is the total dc linked voltage defined in Chap. 2 and v_{dc} is defined total dc linked voltage in Chap. 3 respectively.

$$
\begin{cases}
v_{inva} = S_a v_{dc}/2 \\
v_{invb} = S_b v_{dc}/2 \\
v_{invc} = S_c v_{dc}/2 \\
v_{invg} = S_g v_{dc}/2
\end{cases}
\tag{3.3}
$$

The relation of current and voltage of the DC capacitors are described as:

$$i_{dc} = C_{dc} \frac{dv_{dc}}{dt} \tag{3.4}$$

Following results can be deduced according to (3.4).

$$C \frac{dv_{dc}}{dt} = S_a i_{ca} + S_b i_{cb} + S_c i_{cc} - S_g i_{cn} \tag{3.5}$$

According to (3.1) to (3.5), (3.6) is deduced.

$$\begin{cases} L_c \frac{di_{ca}}{dt} = v_a - S_a v_{dc}/2 \\ L_c \frac{di_{cb}}{dt} = v_b - S_b v_{dc}/2 \\ L_c \frac{di_{cc}}{dt} = v_c - S_c v_{dc}/2 \\ L_{cn} \frac{di_{cn}}{dt} = -v_n + S_g v_{dc}/2 \\ C_{dc} \frac{dv_{dc}}{dt} = \left(S_a i_{ca} + S_b i_{cb} + S_c i_{cc} - S_g i_{cn} \right) \end{cases} \tag{3.6}$$

A general mathematical model of the two-level four-leg VSI in a three-phase four-wire system is established by state space model.

$$Z\dot{X} = AX + BU \tag{3.7}$$

where

$$A = \begin{bmatrix} 0 & 0 & 0 & 0 & -S_a/2 \\ 0 & 0 & 0 & 0 & -S_b/2 \\ 0 & 0 & 0 & 0 & -S_c/2 \\ 0 & 0 & 0 & 0 & S_g/2 \\ S_a & S_b & S_c & -S_g & 0 \end{bmatrix}$$

$$X = \begin{bmatrix} i_{ca} & i_{cb} & i_{cc} & i_{cn} & v_{dc} \end{bmatrix}^T$$
$$B = diag \begin{bmatrix} 1 & 1 & 1 & -1 & 0 \end{bmatrix}$$
$$U = \begin{bmatrix} v_a & v_b & v_c & v_n & 0 \end{bmatrix}$$
$$Z = diag \begin{bmatrix} L_c & L_c & L_c & L_{cn} & C_{dc} \end{bmatrix}$$

3.2.2 Voltage Control Signals According to the Required Compensating Currents

The required compensating currents given by a power quality compensator are discussed in Sect. 2.2. The parallel power quality compensator is operated as a current source, injecting the same but negative amplitude of harmonic, reactive and unbalance currents so that the source can provide balance sinusoidal currents to loads in order to optimize the power flow capacity and reduce loss. It is because the inverter itself is a voltage source inverter but operating as a current source, the inverter should generate the corresponding inverter terminal voltages according to the required compensating currents. According to Fig. 3.10, the voltage vector generated by the voltage source inverter can be given in (3.8).

$$\vec{v}_{inv} = -L_c \frac{d\vec{i}_c}{dt} + \vec{v} \tag{3.8}$$

When dt = T_s, assuming the sampling instants between KT_s and $(K + 1)T_s$, where K is any integer number and T_s is the sampling time, the discrete required compensating voltage, $\vec{v}_{ref}$, from the VSI can be obtained as shown in (3.9). One should be reminded that the vector in this section can be expressed in abc and/or $\alpha\beta$ 0 frames, but both coordinates are equivalent to one another.

$$\vec{v}_{ref}[KT_s] = -\frac{L_c}{T_s}\left\{\vec{i}_c[(K+1)T_s] - \vec{i}_c[KT_s]\right\} + \vec{v}[KT_s] \tag{3.9}$$

When the internal resistance of the coupling inductance can be ignored, (3.9) can be employed. The measured parameters are $\vec{i}_c[KT_s]$ and $\vec{v}[KT_s]$, where $\vec{i}_c[KT_s]$ is the compensating current from the VSI at the time KT_s and $\vec{v}[KT_s]$ is terminal voltage at PCC. The sampling period is already fixed with the value of T_s. The next expected compensating current from the VSI should be $\vec{i}_c[(K+1)T_s]$. If the internal resistance, R_c, of the coupling impedance between the VSI and point of the common coupling terminals cannot be ignored [10], the discrete required compensating voltage, $\vec{v}_{ref}$ from the VSI can be expressed in (3.10).

$$\vec{v}_{ref}[KT_s] = -\frac{R_c}{\Delta X}\left\{\Delta\vec{i}_c[(K+1)T_s] - (1 - \Delta X)\vec{i}_c[KT_s]\right\} + \vec{v}[KT_s] \tag{3.10}$$

where $\Delta X = \frac{R_c}{L_c} \cdot e^{-(R_c/L_c)T_s} \cdot T_s$

3.2.3 Space Vector Analysis of a Four-Leg VSI

In a conventional two-level three-leg VSI as shown in Fig. 3.6, there are $2^3 = 8$ vectors. The constraint is $v_0 = 0$ for any combination of switching functions, S_a, S_b and S_c, as given in (2.55). As a result, the voltage vectors generated by the three-leg VSI are all laid on the $\alpha\beta$ plane. The voltage vector of a three-leg VSI can be given as (3.11), where the switching function can be $S_x = \{+1, -1\}$ and $x = \{a, b, c\}$.

$$\vec{v}_{three-leg} = \frac{v_{dc}}{2}\left(\sqrt{\frac{2}{3}}\left(S_a - \frac{1}{2}S_b - \frac{1}{2}S_c\right) \cdot \vec{n}_\alpha + \frac{1}{\sqrt{2}}(S_b - S_c) \cdot \vec{n}_\beta\right) \tag{3.11}$$

Fig. 3.11 Space vectors of a
two-leg three-leg VSI

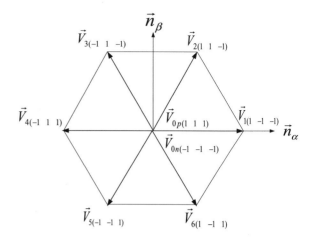

$$\left\| \vec{v}_{three-leg} \right\| = \rho \cdot v_{dc} = r \cdot v_{dc} \quad where \quad \rho = r = 0.816 \qquad (3.12)$$

By considering rectangular and cylindrical coordinates, the parameters, ρ and r, defined in (2.65) and (2.66) are employed. The values of ρ and r in (3.12) are 0.816. Comparing with Table 2.2, ρ and r = 2 × 0.816, it is because the dc voltage of the converter in Chap. 2 is $2v_{dc} = v_{dc1} + v_{dc2}$ but in Chap. 3 dc voltage of the converter is v_{dc} so that (3.12) with $\rho = r = 0.816$. All the angles among directional vectors are 60° apart from one another. There are 2 zero vectors at the origin. Those zero vectors can be employed in arranging the pulse patterns in pulse width modulations to reduce the switching frequency and switching loss. Figure 3.11 shows the space vectors of the three-leg inverter on $\alpha\beta$ plane.

In a two-level four-leg VSI, there are $2^4 = 16$ available vectors. Figure 3.12 shows the projection of the space vectors allocation of four-leg inverter on $\alpha\beta$ plane.

Fig. 3.12 Space vectors of a
two-level four-leg VSI on $\alpha\beta$
plane

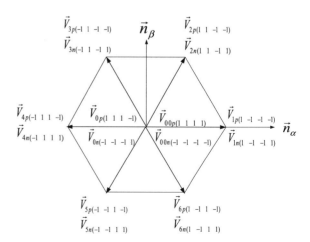

Comparing space vectors of three-leg and four-leg inverters on $\alpha\beta$ plane only, one can find that all vectors, except zero vectors, have the same length of $r * v_{dc}$, in which $r = 0.816$. Table 3.1 lists all the vectors corresponding to their switching functions.

Figure 3.13 shows all space vectors of a four-leg VSI in three-dimension. Table 3.1 lists all the vectors from abc into $\alpha\beta0$ coordinates, where v_{dc} is the instantaneous dc link voltage of the four-leg VSI. Table 2.4 lists all voltage vectors when $2v_{dc}$ is the instantaneous dc link voltage. One can find that these 2 tables have a factor 0.5 difference. Those vector characteristics in three-dimension can be classified and considered as follows when v_{dc} is the instantaneous dc link voltage:

(1) Vectors, $\vec{V}_{00n}$ and $\vec{V}_{00p}$, are perfect zero vectors which can be used for arranging the pulse patterns to reduce the switching frequency and switching loss. Only $\vec{V}_{00n}$ and $\vec{V}_{00p}$ are laid on the origin of $\alpha\beta$ plane.
(2) Vectors, $\vec{V}_{0n}$ and $\vec{V}_{0p}$, are zero vectors. However, both of them can inject only zero component voltage into the system and they do not have any voltage

Table 3.1 Two-level four-leg VSI voltage vectors

	S_f	S_a	S_b	S_c	V_{ag}	V_{bg}	V_{cg}	S_α	S_β	S_0	V_α	V_β	V_0
$\vec{V}_{1p}$	−1	1	−1	−1	v_{dc}	0	0	2	0	2	$\sqrt{\frac{2}{3}}v_{dc}$	0	$\frac{1}{\sqrt{3}}v_{dc}$
$\vec{V}_{2p}$	−1	1	1	−1	v_{dc}	v_{dc}	0	1	2	4	$\sqrt{\frac{1}{6}}v_{dc}$	$\frac{1}{\sqrt{2}}v_{dc}$	$\frac{2}{\sqrt{3}}v_{dc}$
$\vec{V}_{3p}$	−1	−1	1	−1	0	v_{dc}	0	−1	2	2	$-\sqrt{\frac{1}{6}}v_{dc}$	$\frac{1}{\sqrt{2}}v_{dc}$	$\frac{1}{\sqrt{3}}v_{dc}$
$\vec{V}_{4p}$	−1	−1	1	1	0	v_{dc}	v_{dc}	−2	0	4	$-\sqrt{\frac{2}{3}}v_{dc}$	0	$\frac{2}{\sqrt{3}}v_{dc}$
$\vec{V}_{5p}$	−1	−1	−1	1	0	0	v_{dc}	−1	−2	2	$-\sqrt{\frac{1}{6}}v_{dc}$	$\frac{-1}{\sqrt{2}}v_{dc}$	$\frac{1}{\sqrt{3}}v_{dc}$
$\vec{V}_{6p}$	−1	1	−1	1	v_{dc}	0	v_{dc}	1	−2	4	$\sqrt{\frac{1}{6}}v_{dc}$	$\frac{-1}{\sqrt{2}}v_{dc}$	$\frac{2}{\sqrt{3}}v_{dc}$
$\vec{V}_{0p}$	−1	1	1	1	v_{dc}	v_{dc}	v_{dc}	0	0	6	0	0	$\sqrt{3}v_{dc}$
$\vec{V}_{00n}$	−1	−1	−1	−1	0	0	0	0	0	0	0	0	0
$\vec{V}_{1n}$	1	1	−1	−1	0	v_{dc}	v_{dc}	2	0	−4	$\sqrt{\frac{2}{3}}v_{dc}$	0	$-\frac{2}{\sqrt{3}}v_{dc}$
$\vec{V}_{2n}$	1	1	1	−1	0	0	$-v_{dc}$	1	2	−2	$\sqrt{\frac{1}{6}}v_{dc}$	$\frac{1}{\sqrt{2}}v_{dc}$	$-\frac{1}{\sqrt{3}}v_{dc}$
$\vec{V}_{3n}$	1	−1	1	−1	$-v_{dc}$	0	$-v_{dc}$	−1	2	−4	$-\sqrt{\frac{1}{6}}v_{dc}$	$\frac{1}{\sqrt{2}}v_{dc}$	$-\frac{2}{\sqrt{3}}v_{dc}$
$\vec{V}_{4n}$	1	−1	1	1	$-v_{dc}$	0	0	−2	0	−2	$-\sqrt{\frac{2}{3}}v_{dc}$	0	$-\frac{1}{\sqrt{3}}v_{dc}$
$\vec{V}_{5n}$	1	−1	−1	1	$-v_{dc}$	$-v_{dc}$	0	−1	−2	−4	$-\sqrt{\frac{1}{6}}v_{dc}$	$\frac{-1}{\sqrt{2}}v_{dc}$	$-\frac{2}{\sqrt{3}}v_{dc}$
$\vec{V}_{6n}$	1	1	−1	1	0	$-v_{dc}$	0	1	−2	−2	$\sqrt{\frac{1}{6}}v_{dc}$	$\frac{-1}{\sqrt{2}}v_{dc}$	$-\frac{1}{\sqrt{3}}v_{dc}$
$\vec{V}_{0n}$	1	−1	−1	−1	$-v_{dc}$	$-v_{dc}$	$-v_{dc}$	0	0	−6	0	0	$-\sqrt{3}v_{dc}$
$\vec{V}_{00p}$	1	1	1	1	0	0	0	0	0	0	0	0	0

Fig. 3.13 Space vectors of a two-level four-leg VSI in three-dimension

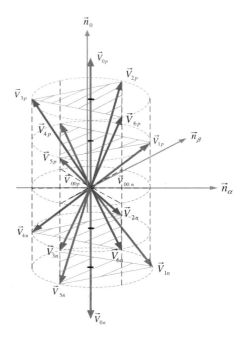

component on $\alpha\beta$ plane. Its parameters in zero components are $\rho = Z_0 = 1.732$.

(3) When $Z_0 = 1.155$, there are vectors, $\vec{V}_{1n}, \vec{V}_{2p}, \vec{V}_{3n}, \vec{V}_{4p}, \vec{V}_{5n}, \vec{V}_{6p}$, in which $\vec{V}_{2p}, \vec{V}_{4p}$ and $\vec{V}_{6p}$ are laid on $Z_0 = 1.155$. On the other hand, $\vec{V}_{1n}, \vec{V}_{3n}$ and $\vec{V}_{5n}$ are laid on $Z_0 = -1.155$. The norm of those vectors is $\rho = 1.414$.

(4) When $Z_0 = \pm 0.557$, there are vectors, $\vec{V}_{1p}, \vec{V}_{2n}, \vec{V}_{3p}, \vec{V}_{4n}, \vec{V}_{5p}, \vec{V}_{6n}$, in which $\vec{V}_{1p}, \vec{V}_{3p}$ and $\vec{V}_{5p}$ are laid on $Z_0 = 0.557$. On the other hand, $\vec{V}_{2n}, \vec{V}_{4n}$ and $\vec{V}_{6n}$ are laid on $Z_0 = -0.557$. The norm of those vectors is $\rho = 1$.

(5) When $Z_0 > 0$ is required, vectors, $\vec{V}_{1p}, \vec{V}_{2p}, \vec{V}_{3p}, \vec{V}_{4p}, \vec{V}_{5p}, \vec{V}_{6p}$ and $\vec{V}_{0p}$, can be chosen. And vice versa, $\vec{V}_{1n}, \vec{V}_{2n}, \vec{V}_{3n}, \vec{V}_{4n}, \vec{V}_{5n}, \vec{V}_{6n}$ and $\vec{V}_{0n}$ can be chosen for $Z_0 < 0$.

(6) Finally, its output voltage of the four-leg VSI can be expressed as (3.13). All vectors, except zero vectors ($\vec{V}_{0n}, \vec{V}_{0p}, \vec{V}_{00n}$ and $\vec{V}_{00p}$), have r = 0.816 on $\alpha\beta$ plane.

$$\vec{v}_{four-leg} = \frac{v_{dc}}{2}\left(\sqrt{\frac{2}{3}}\left(S_a - \frac{1}{2}S_b - \frac{1}{2}S_{cb} \right) \cdot \vec{n}_\alpha + \frac{1}{\sqrt{2}}(S_b - S_c) \cdot \vec{n}_\beta \right.$$
$$\left. + \frac{1}{\sqrt{3}}(S_a + S_b + S_c - 3S_g) \cdot \vec{n}_0 \right)$$

(3.13)

3.2.4 Hysteresis PWM

The pulse width modulation plays an important role in the APF control system to inject the required current for compensation. It generates the trigger signals for each power electronics switches in terms of the compensating current references. In this section, the hysteresis PWM method for two-level four-leg inverters is introduced.

3.2.4.1 Basic Hysteresis PWM

The basic operation principle of hysteresis pulse width modulation (PWM) can be explained in Fig. 3.14, where i_{cx}^* is the reference tracking current and i_{cx} is the actual injected current. There are upper and lower margins between i_{cx}^*. When the actual current i_{cx} reaches the upper margin of i_{cx}^*, the switching state is changed to negative state. Vice Versa, when the i_{cx} reaches the lower margin of i_{cx}^*, then the switching state is changed to positive. As a result, the actual current can be oscillating somewhere between the upper and lower margins of the reference tracking current, in which the VSI actually operates as a controlled current source.

3.2.4.2 Hysteresis PWM for Two-Level Four-Leg VSI

The block diagram of the hysteresis PWM for the two-level four-leg VSI is shown in Fig. 3.15. The current references are compared with the measured output currents of the VSI. The obtained error passes through the hysteresis controller and then the trigger signals for the switching devices are generated. The switching frequency is not fixed [11, 12] and it varies according to the error band and system parameters. By using instantaneous reactive power theory [13–15] and mixed coordinate instantaneous compensation method described in Sect. 2.2, the reference tracking current can be calculated.

In *a-b-c-0* coordinates, the upper and lower error bands or tolerances of hysteresis methods can be defined as following.

Fig. 3.14 Generation of trigger signals of the hysteresis PWM

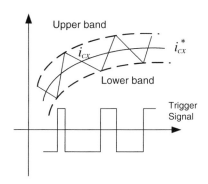

Fig. 3.15 Hysteresis PWM
for two-level four-leg VSI

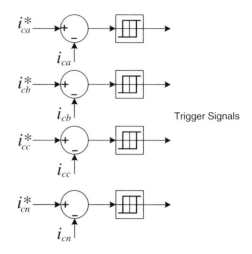

$$\text{Upper Band Limit} : i_{cx}^* + B$$
$$\text{Lower Band Limit} : i_{cx}^* - B$$

where i_{cx}^* (k = a, b, c, n) is the reference tracking current and B is the total error-allowable band width. The error bands in a-, b- and c-axis form a cubic in a-b-c coordinates, as shown in Fig. 3.16. And the volume of the error cubic is equal to $8B^3$.

The error in the fourth-leg of the two-level four-leg VSI should be considered in a PWM method. The error band of the current passing through the fourth-leg, i.e., the error of the neutral current compensation, varies between $-B$ and B. However, when the two-level four-leg VSI is used as a shunt power quality compensator, as shown in Fig. 3.10, the error of the fourth-leg and the other three-leg output are not independent of each other.

Fig. 3.16 Error volume in
a-b-c coordinates

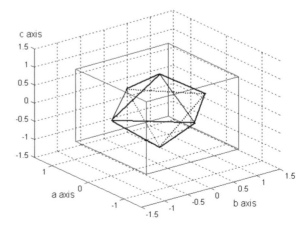

The following relations should exist: $i_{ca} + i_{cb} + i_{cc} = i_{cn}, i_{cx} = i_{cx}^* + e_x$, where i_{cx} (x = a, b, c, n) represents the output currents of the VSI, i_{cx}^* represents the current reference and e_x (x = a, b, c, n) represents the compensation error. Since $i_{ca}^* + i_{cb}^* + i_{cc}^* = i_{cn}^*$, the following result is derived: $e_a + e_b + e_c = e_n$. The hysteresis band of the fourth-leg includes an extra restrict to the total error band in the a-b-c coordinates, i.e., $-B \leq e_a + e_b + e_c = e_n \leq B$. After this new limits are applied, the error cubic in the a-b-c coordinates is shrunk to an octahedron, as shown in Fig. 3.16. The volume of the error octahedron is $4B^3/3$. Consequently, when compared with the results from [16], the two-level four-leg VSI provides a better performance compared to that of the two-level center-split VSI when the hysteresis PWM is used, especially in the neutral current compensation.

3.2.4.3 Simulation Results

The simulated system configuration is shown in Fig. 3.10 when the two-level four-leg VSI is applied to a three-phase four-wire APF. The source currents are shown in Fig. 3.17. The APF began to operate at 0.04 s. The THD of the source currents is reduced from 39.9 to 5.4 %. The neutral currents drop from 12.1 to 1.7 A. The sampling frequency is 10 kHz, that is to say, the trigger signals are determined in every 100 us by comparing current tracking error and the error band. In principle, the switching frequency is not a fixed value in the hysteresis PWM as it

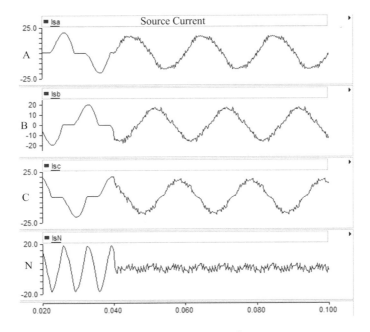

Fig. 3.17 Source currents before and after APF compensation

depends on error band, coupling inductance and instantaneous voltage difference between the point of common coupling terminal voltage and VSI terminal voltage. By fixing the sampling frequency, the error band may be larger than the predefined value. The recorded switching frequency is around 2.5 kHz for this simulation.

As discussed in the previous sections, the *a-b-c-0* hysteresis PWM control of the two-level four-leg VSI includes an extra constraint for the error in zero-sequence. As a result, the neutral current compensation performance is better when a two-level four-leg VSI is used. Results of the comparison with the three-leg center-split VSI are given in the following sections.

3.2.5 Space Vector Modulation

3.2.5.1 Basic Space Vector Modulation for Two-Level Three-Leg VSI

In this section, a basic two-level three-leg VSI is discussed and basic concept of space vector modulation (SVM) is introduced in 2-dimension $\alpha\beta$ plane. Another approach of PWMs can be based on the space vector representation of the voltages on $\alpha\beta$ plane. There are eight states available according to eight switching positions of the conventional three-leg VSI as depicted in Fig. 3.18 . The space can be further divided into totally 6 sectors as shown in Fig. 3.18. The required reference voltage, $\vec{V}_{ref}$, is assumed constant during one switching cycle.

In Fig. 3.19, the reference vector, $\vec{V}_{ref}$ is located in sector I. The voltage-second reference can be approximated by a sequence of voltage-second states. Hence, (3.14) can be given for a switching cycle in sector I and (3.15) is a constraint for keeping the switching time to be T_s.

Fig. 3.18 Space vectors of a conventional three-leg VSI

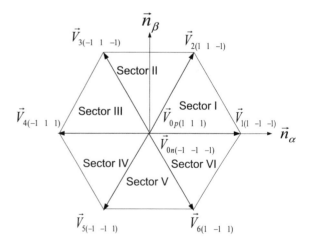

Fig. 3.19 Determination of switching times

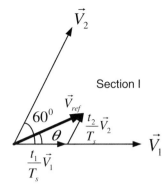

$$\vec{V}_{ref} T_s = \vec{V}_1 t_1 + \vec{V}_2 t_2 + \vec{V}_0 t_0 \tag{3.14}$$

$$T_s = t_1 + t_2 + t_0 \tag{3.15}$$

Referring to Fig. 3.6, the voltage vectors $\vec{V}_{1(1,-1,-1)}$ and $\vec{V}_{2(1,1,-1)}$, can be expressed as:

$$\vec{V}_{1(1,-1,-1)} = \sqrt{\frac{2}{3}} v_{dc} \cdot \vec{n}_\alpha \tag{3.16}$$

$$\vec{V}_{2(1,1,-1)} = \sqrt{\frac{2}{3}} \cdot v_{dc} \left(\frac{1}{2} \cdot \vec{n}_\alpha + \frac{\sqrt{3}}{2} \cdot \vec{n}_\beta \right) \tag{3.17}$$

Furthermore, when the space vectors are described in rectangular coordinates, it follows that:

$$\vec{V}_{ref} \begin{bmatrix} \cos\theta \\ \sin\theta \end{bmatrix} T_s = \sqrt{\frac{2}{3}} v_{dc} \begin{bmatrix} 1 \\ 0 \end{bmatrix} t_1 + \sqrt{\frac{2}{3}} v_{dc} \begin{bmatrix} \cos 60° \\ \sin 60° \end{bmatrix} t_2 \tag{3.18}$$

Hence,

$$t_1 = T_s \cdot \frac{|\vec{V}_{ref}|}{\sqrt{\frac{2}{3}} v_{dc}} \cdot \frac{\sin(60° - \theta)}{\sin 60°} \tag{3.19}$$

$$t_2 = T_s \cdot \frac{|\vec{V}_{ref}|}{\sqrt{\frac{2}{3}} v_{dc}} \cdot \frac{\sin\theta}{\sin 60°} \tag{3.20}$$

$$t_0 = T_s - t_1 - t_2 \tag{3.21}$$

The zero vector switching time is t_0 and zero vectors are $\vec{V}_{0p(1,1,1)}$ and $\vec{V}_{0n(-1,-1,-1)}$. One of them can be chosen to reduce the switching frequency. To obtain minimum switching frequency of each inverter leg, it is necessary to arrange the switching sequence in such a way that the transition from one state to the next is performed by switching only one inverter leg [17]. It can be seen in Fig. 3.20 that the above conditions are met if starting from one zero state, the inverter legs are switched in the relevant sequence ending at the other zero state. Similarly, the above approach can be applied in other sections. An optimum pulse pattern of space vector PWM is given in Fig. 3.20, in which the switching sequence is $\vec{V}_{0n}\vec{V}_1\vec{V}_2\vec{V}_{0p}\vec{V}_2\vec{V}_1\vec{V}_{0n}$.

Hereafter, the space vector modulation for four-leg VSI is explained and it can be considered in three-dimension space and this approach can be classified as 3D SVM in four-leg VSI.

3.2.5.2 Space Vector Modulation for Two-Level Four-Leg VSI

The 3DSVM proposed in [18] is to calculate the switching times with respective nodes a, b, c, and n as shown in Fig. 3.10, which has four steps: (1) Prism Identification, (2) Tetrahedron Selection, (3) Duty Ratio Calculation of each non-zero voltage vectors and (4) Switching Sequences Determination. Those steps are explained as follows. However, by using 3DSVM to control the inverter to

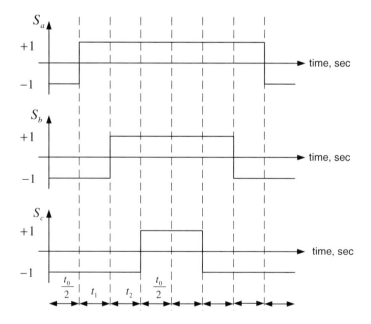

Fig. 3.20 Optimum pulse pattern of space vector modulation (SVM)

operate as an active power filter, the reference voltage generated by 3DSVM should create the corresponding compensating reference current accordingly. By using the mixed coordinate instantaneous compensation method described in Sect. 2.2, the reference tracking current can be calculated. However, the corresponding voltage vectors should be calculated according to Sect. 3.2.2 so that the VSI can be operated as a current source to inject the required compensating current.

A traditional three-leg VSI has eight possible switching combinations. With the additional fourth leg, the total number of switching combinations increases to sixteen. The 2^4 switching vectors can be displayed in $\alpha\beta0$ frame as shown in Fig. 3.13.

Step (1) Prism Identification

According to the defined parameters, θ and ϕ, as shown in Fig. 3.21, these parameters of the reference voltage vectors are calculated. The Prism sections [18] can be classified as given in Table 3.2, which are the same definition as the Section I, II, III, IV, V and VI as shown in Fig. 3.18. Finally, a Prism can be decided based on the projections of the reference voltage vector on $\alpha\beta$ plane.

Step (2) Tetrahedron Selection

In order to minimize the current ripple, switching vectors adjacent to the reference vector should be selected. Within the selected prism, there are six non-zero switching vectors and 2 zero switching vectors, which can further divide the prism into four tetrahedrons. For example, when the reference voltage is inside Prism I, the reference vector can be within:

Tetrahedron I which defined and bounded by $\vec{V}_{1p}, \vec{V}_{1n}, \vec{V}_{2n}, \vec{V}_{00p}$ and $\vec{V}_{00n}$
Tetrahedron II which defined and bounded by $\vec{V}_{1p}, \vec{V}_{2n}, \vec{V}_{2p}, \vec{V}_{00p}$ and $\vec{V}_{00n}$
Tetrahedron III which defined and bounded by $\vec{V}_{0p}, \vec{V}_{1p}, \vec{V}_{2p}, \vec{V}_{00p}$ and $\vec{V}_{00n}$
Tetrahedron IV which defined and bounded by $\vec{V}_{0n}, \vec{V}_{1n}, \vec{V}_{2n}, \vec{V}_{00p}$ and $\vec{V}_{00n}$

Fig. 3.21 Defined parameters in coordinates

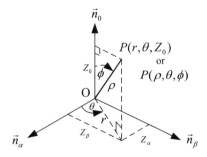

Table 3.2 Prism classification

Prism I	Prism II	Prism III	Prism IV	Prism V	Prism VI
$0 \le \theta \le 60°$	$60° \le \theta \le 120°$	$120° \le \theta \le 180°$	$180° \le \theta \le 240°$	$240° \le \theta \le 300°$	$300° \le \theta \le 360°$

For simplicity, Fig. 3.22 shows those six non-zero directional vectors, $\vec{V}_{0p}, \vec{V}_{0n}, \vec{V}_{1p}, \vec{V}_{1n}, \vec{V}_{2p}$ and $\vec{V}_{2n}$, as well as two zero vectors $\vec{V}_{00p}$ and $\vec{V}_{00n}$ for Prism I in $\alpha\beta0$ frame. Each tetrahedron has three non-zero directional vectors and two zero vectors and they all are adjacent to each other.

Synthesizing the reference vector by using adjacent vectors inside a tetrahedron of a prism minimizes the current ripple. The tetrahedron identification can be achieved by referring the voltage polarities of the reference vector in abc coordinate. The tetrahedron that encloses the reference vector has the same voltage polarity set. The selection of zero vectors on the sequencing scheme can provide a degree of freedom to reduce the switching loss and harmonic component generated.

Step (3)   Calculation of Switching Times

Referring to (3.14) and (3.15) for 2 dimensional SVM, the switching times for adjacent vectors in 3D SVM can be calculated by:

$$\vec{V}_{ref}T_s = \vec{V}_1 t_1 + \vec{V}_2 t_2 + \vec{V}_3 t_3 + \vec{V}_0 t_0 \tag{3.22}$$

Fig. 3.22 Six non-zero directional and two zero vectors

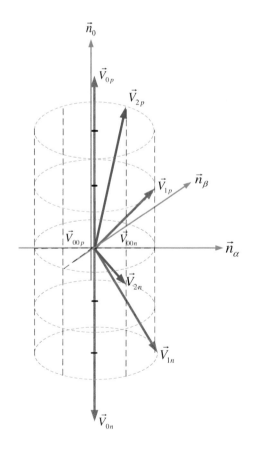

$$T_s = t_1 + t_2 + t_3 + t_0 \tag{3.23}$$

For example, the reference vector is allocated in Tetrahedron I of Prism I so that adjacent vectors, $\vec{V}_{1p}, \vec{V}_{1n}$ and $\vec{V}_{2n}$, are selected in order to synthesize the reference vector by using voltage-second concept. In this case, $\vec{V}_{1p} = \vec{V}_1$, $\vec{V}_{1n} = \vec{V}_2$ and $\vec{V}_{2n} = \vec{V}_3$ are selected. Then, $\vec{V}_0$ can either be $\vec{V}_{00p}$ or $\vec{V}_{00n}$. According to different locations of the reference voltage vector, $\vec{V}_1, \vec{V}_2, \vec{V}_3$ and $\vec{V}_0$ are selected correspondingly. There are totally 24 sectors (6 Prisms and each of them has 4 Tetrahedron). Therefore, the pre-calculated 24-matrix table is required in practical implementation.

Step (4)   Switching Sequences

The ON time calculation of each switch is based on the arrangements of non-zero voltage vectors and zero vectors. However, the distribution of t_0 on zero vector(s), $\vec{V}_{00p}$ or/and $\vec{V}_{00n}$, can be chosen based on switching loss and/or harmonic current distortion. The detailed switching sequences and comparison among them are not included in this book but that information can be founded in [18].

3.2.5.3   3D Direct PWM for Four-Leg Inverters

In above section, a 3DSVM is explained. However, it requires several steps for implementation of 3DSVM. Therefore, the 3DSVM of a two-level four-leg VSI involves complex calculations and switching tables. Relatively, comparing this 3DSVM with a carrier based PWM method, one can find that a carrier based PWM [19] is equivalent to the above 3DSVM, but its implementation is easier. The basic idea of the carrier based PWM is to generate the PWM outputs by direct determination of the reference voltages for each leg of a four leg VSI. The 3D direct PWM [20] is also based on generating the pulse of each leg independently. Therefore, the 3D direct PWM can be used to control a four leg VSI, which is given in this section.

The direct PWM is based on the volt-second approximation just like the conventional SVM. The output voltage of each leg is synthesized by two voltage levels for a two-level VSI. The single leg output model is given in the Fig. 3.23. A virtual ground is introduced in the middle of the dc bus. The dc capacitor is split to two, but actually one dc capacitor can be used. If the target output of this leg is V_{refx} (x = a, b, c, g) and the period for the pulse width modulation is T_s, the approximation of the output voltage is expressed in (3.24), in which T_{onx} (x = a, b, c, g) is the pulse width when the upper switch is turned on and the lower switch is turned off. During this time interval, the output voltage V_{invx} (x = a, b, c, g) equals to $V_{dc}/2$. In the remained time of one period, the upper switch is off and the lower switch is on and V_{invx} equals to $-V_{dc}/2$.

Fig. 3.23 Single leg output model of two-level VSI

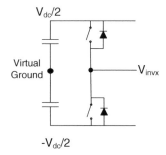

$$V_{refx}T_s = V_{dc}/2 \cdot T_{onx} - V_{dc}/2 \cdot (T_s - T_{onx}) \tag{3.24}$$

From (3.24), (3.25) is deduced. By normalizing the reference voltage with the dc-link voltage and normalizing the pulse width with the period T_s, (3.26) can be further deduced. It indicates that the pulse width of each leg can be directly calculated by the normalized reference voltage.

$$T_{onx}/T_S = V_{refx}/V_{dc} + 1/2 \tag{3.25}$$

$$t_{onx} = v_{refx} + 1/2 \tag{3.26}$$

The output phase-to-neutral voltages of two-level four-leg VSIs are determined by the difference between the output voltages of inverter leg x ($x = a, b, c$) and the fourth-leg g.

$$v_{xg} = v_{refx} - v_{refg} \quad x = a, b, c \tag{3.27}$$

According to (3.26), the normalized pulse width of legs A, B and C are expressed as (3.28) and the normalized pulse width of the fourth leg is expressed as (3.29).

$$t_{onx} = v_{xg} + v_{refg} + 1/2 \quad x = a, b, c \tag{3.28}$$

$$t_{ong} = v_{refg} + 1/2 \tag{3.29}$$

The v_{xg} ($x = a, b, c$) is the phase-to-neutral reference voltage for controlling four-leg VSIs. If a parameter v_{gg} is introduced, which always equals to zero, (3.28) can also be used to calculate the pulse width of the fourth leg. The corresponding output pulses of a two-level four-leg inverter are shown in Fig. 3.24. The rising-edge aligned scheme was chosen for the convenience in illustrating the effective time in the following discussions.

The effective time is defined as (3.30), which equals the difference between the dwell times of legs A, B or C and the fourth leg g, as shown in Fig. 3.24. If (3.28) and (3.29) are substituted by (3.30), (3.31) indicates that output phase-to-neutral

Fig. 3.24 Output pulse
widths and effective times

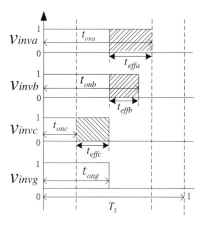

voltages are generated in dwell time t_{effx} ($x = a, b, c$). In addition, the effective time
t_{effx} is equal to the normalized phase-to-neutral voltage.

$$t_{effx} = t_{onx} - t_{ong} \quad x = a, b, c \tag{3.30}$$

$$t_{effx} = t_{onx} - t_{ong} = v_{xg} + v_{refg} + 1/2 - (v_{refg} + 1/2) = v_{xg} \tag{3.31}$$

$$0 < v_{xf} + v_{refg} + 1/2 < 1 \; x = a, b, c, g \tag{3.32}$$

Only the reference voltage v_{refg} of the fourth-leg is unknown when the pulse
width of each leg is calculated by (3.28). According to (3.31), the variation of the
v_{refg} does not affect the output phase-to-neutral voltages of a four-leg VSI. Hence,
the v_{refg} can be freely chosen if the condition in (3.32) is satisfied, which limits the
normalized pulse widths in the range of 0–1.

(3.32) can be rewritten as:

$$-1/2 - v_{min} < v_{refg} < 1/2 - v_{max} \tag{3.33}$$

where v_{max} is defined as the maximum number from among v_{ag}, v_{bg}, v_{cg} and v_{gg}; and
v_{min} is the minimum number from among v_{ag}, v_{bg}, v_{cg} and v_{gg}. (3.33) only provides
the range for v_{refg}. In order to get a fixed value for v_{refg}, an extra constraint is
adopted.

Besides the effective time for generating output voltages, there are redundant
times in one PWM period which are taken up by the switching states (0, 0, 0, 0) and
(1, 1, 1, 1). The output phase-to-neutral voltages are zero when the four-leg VSI
operates according to the two switching patterns. It is assumed that the dwell times
of states (0, 0, 0, 0) and (1, 1, 1, 1) are the same, so that the optimal switching
sequence can be achieved. Consequently, (3.34) is obtained.

$$t_{on\,min} = 1 - t_{on\,max,} \tag{3.34}$$

where $t_{on\,min}$ is the dwell time of switching states (1, 1, 1, 1), which is also the minimum pulse width among the four legs, as illustrated in Fig. 3.24; $1 - t_{on\,max}$ is the dwell time of switching states (0, 0, 0, 0). If (3.28) is considered, (3.35) and (3.36) can be obtained. (3.37) is obtained by substituting (3.35) and (3.36) into (3.34).

$$t_{on\,min} = v_{min} + v_{refg} + 1/2 \tag{3.35}$$

$$t_{on\,max} = v_{max} + v_{refg} + 1/2 \tag{3.36}$$

$$v_{refg} = -(v_{max} + v_{min})/2 \tag{3.37}$$

Therefore, the shifting voltage is defined as:

$$v_{shift} = -(v_{max} + v_{min})/2 \tag{3.38}$$

where v_{max} is the maximum number from among v_{ag}, v_{bg}, v_{cg} and v_{gg}; v_{min} is the minimum number from among v_{ag}, v_{bg}, v_{cg} and v_{gg}.

The reference voltage of each leg of a two-level four-leg VSI can be calculated once the shifting voltage has been determined.

$$v_{refx} = v_{xg} + v_{shift} \quad x = a, b, c, g \tag{3.39}$$

According to (3.28), the output pulses can be generated using the pulse widths determined by (3.40) for a two-level four-leg VSI.

$$t_{onx} = v_{xg} + v_{shift} + 1/2 \quad x = a, b, c, g \tag{3.40}$$

3.2.5.4   Simulation Results

The system configuration as shown in Fig. 3.10 is also used to verify the validity of the direct PWM for the two-level four-leg VSI. One should be reminded that 3DSVM is equivalent to a carried based PWM and it can be implemented by direct PWM. As a result, only direct PWM is simulated in Table 3.3. The source currents are shown in Fig. 3.25. The APF began to operate at 0.04 s. The THD of the source currents reduce from 39.9 to around 5 %. The neutral current drops from 12.1 to 1.65 A. The sampling frequency is 5 kHz, just half of that required by a hysteresis PWM. When Fig. 3.25 is compared to Fig. 3.17, the results indicate that the current ripple can be greatly reduced. The detailed comparison of the compensation performance is provided in Table 3.3.

Table 3.3 Comparison of compensation performance of four-leg VSI

	THDA (%)	THDB (%)	THDC (%)	Neutral currents (A)
Load currents	39.96	39.96	39.96	12.16
Hysteresis PWM	9.06	9.68	9.27	1.75
Direct PWM	5.6	5.34	5.05	1.65

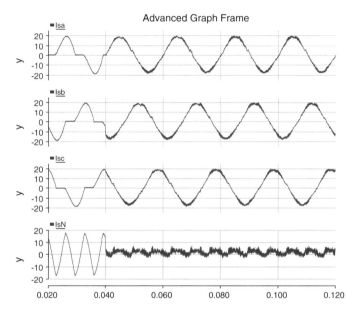

Fig. 3.25 Source currents before and after APF comparison

3.3   Two-Level and Three-Level Three-Leg Center-Split VSI as Three-Phase Four-Wire Active Power Filters

Three-leg center-split voltage source inverters (VSIs) are proposed for applications in three-phase four-wire systems [3, 21–23]. The DC capacitor is split to two, so that the neutral wire of the three-phase four-wire system can be connected to the inverter. The two-level center-split VSI based APF system configuration is shown in Fig. 3.26, which can be operated in a three-phase four-wire power system.

However, for the higher voltage power applications, the multi-level VSI topologies are good alternatives. The multi-level structure not only reduces voltage stress across the power electronics switches but also provides many more available vectors. Therefore, it improves harmonic output contents of the VSI by selecting appropriate switching vectors. The most popular topology is the three-level neutral-point-clamped (NPC) inverter [24, 25]. Since the dc capacitor is split to two, this inverter can be used in a three-phase four-wire APF without modification.

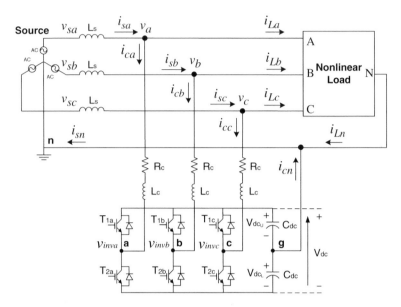

Fig. 3.26 A three-phase four-wire APF uses a two-level center-split VSI

The system configuration of a three-level three-phase four-wire APF is shown in Fig. 3.27. It is obvious that the traditional three-level neutral point clamped inverter can be directly applied into three-phase four-wire applications [26] from the three-phase three-wire systems if and only if a neutral wire is connected as shown in Fig. 3.27.

3.3.1   Modeling of Three-Leg Center-Split Active Power Filters

The modeling of the three-leg center-split active power filters is given in this section. The active power filter can be a two-level or three-level VSI. The equivalent model of the three-level VSI is shown in Fig. 3.28.

From Fig. 3.27, the switching functions are considered as the equivalent switched devices such as IGBT's, e.g. in phase A, S_a can be written as:

$$S_x = \begin{cases} 1, & T_{1x}\,on, \quad T_{2x}\,on, \quad T_{3x}\,off, \quad T_{4x}off \\ 0, & T_{1x}\,off, \quad T_{2x}\,on, \quad T_{3x}\,on, \quad T_{4x}off \\ -1, & T_{1x}\,off, \quad T_{2x}\,off, \quad T_{3x}\,on, \quad T_{4x}off \end{cases} \quad (x = a, b, c) \quad (3.41)$$

There are three cases in one leg of the three-level VSI such as positive, zero or negative switching function in the equivalent model in Fig. 3.3.

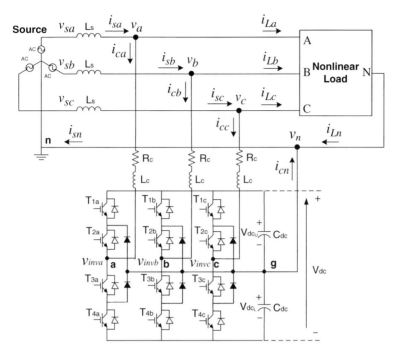

Fig. 3.27 Three-phase four-wire APF uses a three-level three-leg inverter

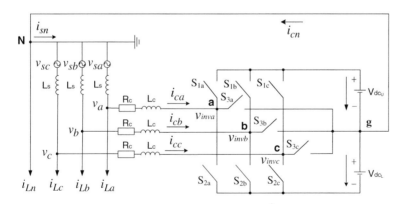

Fig. 3.28 Equivalent model of three-level VSI correct S_{3a} to S_{0a}

(1) If $S_a = 1$, then $S_{1a} = 1$, $S_{2a} = 0$, $S_{3a} = 0$;

(2) If $S_a = 0$, then $S_{1a} = 0$, $S_{2a} = 0$, $S_{3a} = 1$;

(3) If $S_a = -1$, then $S_{1a} = 0$, $S_{2a} = 1$, $S_{3a} = 0$;

It is noticed that S_a, S_b and S_c can be 1, 0 and -1. The boundary condition of S_{1a}, S_{2a} and S_{3a} is defined as

$$\begin{cases} S_{1a} + S_{2a} + S_{3a} = 1 \\ S_{1a} = 1 \ or \ 0, \quad S_{2a} = 1 \ or \ 0, \quad S_{3a} = 1 \ or \ 0 \end{cases} \qquad (3.42)$$

It means that when S_{1a} is equal to 1, then S_{2a} and S_{3a} must be zero. When the two-level VSI is employed, $S_a = 0$ is not existed. As a result, the modelling in this section can be used for both two-level and three-level cases.

According to Fig. 3.28, the relation among the ac-side compensating current i_{cx}, the terminal voltage of the VSI v_{invx}, the terminal voltages at PCC v_x and coupling inductors L_c can be expressed as (3.43), where $x = a, b, c$.

$$\begin{cases} L_c \dfrac{di_{ca}}{dt} = -R_c \cdot i_{ca} - v_{inva} + v_a \\[2mm] L_c \dfrac{di_{cb}}{dt} = -R_c \cdot i_{cb} - v_{invb} + v_b \\[2mm] L_c \dfrac{di_{cc}}{dt} = -R_c \cdot i_{cc} - v_{invc} + v_c \end{cases} \qquad (3.43)$$

By using the switching functions, the relation between the terminal voltage and the dc-link voltage can be expressed as:

$$\begin{cases} v_{inva} = S_{1a} \cdot v_{dcU} - S_{2a} \cdot v_{dcL} \\ v_{invb} = S_{1b} \cdot v_{dcU} - S_{2b} \cdot v_{dcL} \\ v_{invc} = S_{1c} \cdot v_{dcU} - S_{2c} \cdot v_{dcL} \end{cases} \qquad (3.44)$$

A general mathematical model of the three-level converter in a three-phase four-wire system can be established as follows:

$$\mathbf{Z\dot{X}} = \mathbf{AX} + \mathbf{BU} \qquad (3.45)$$

where

$$A = \begin{bmatrix} -R_c & 0 & 0 & -S_{1a} & S_{2a} \\ 0 & -R_c & 0 & -S_{1b} & S_{2b} \\ 0 & 0 & -R_c & -S_{1c} & S_{2c} \end{bmatrix}$$

$$\mathbf{X} = \begin{bmatrix} i_{ca} & i_{cb} & i_{cc} & V_{dcU} & V_{dcL} \end{bmatrix}^T$$
$$\mathbf{B} = diag\begin{bmatrix} 1 & 1 & 1 \end{bmatrix}$$
$$\mathbf{U} = \begin{bmatrix} v_a & v_b & v_c \end{bmatrix}$$
$$\mathbf{Z} = diag\begin{bmatrix} L_c & L_c & L_c \end{bmatrix}$$

3.3.2 Space Vector Analysis of a Three-Leg Center-Split VSI

In a three-phase three-wire or three-phase four-wire system, an instantaneous voltage vector of the VSI, $\vec{v}$, can be represented by a linear combination of the vectors, $\vec{n}_a, \vec{n}_b$ and $\vec{n}_c$, which can form a basis, $B = \{\vec{n}_a, \vec{n}_b, \vec{n}_c\}$, for a vector to span with

$$\vec{v} = K_a\vec{n}_a + K_b\vec{n}_b + K_c\vec{n}_c \qquad (3.46)$$

where K_a, K_b and K_c are scalars.

The three-phase instantaneous voltages of the VSI can be transferred from abc into $\alpha\beta0$ coordinates in three-phase four-wire systems by the coordinate transformation matrix as given in (3.47)

$$\begin{bmatrix} v_\alpha \\ v_\beta \\ v_0 \end{bmatrix} = \sqrt{\frac{2}{3}} \begin{bmatrix} 1 & -1/2 & -1/2 \\ 0 & \sqrt{3}/2 & -\sqrt{3}/2 \\ 1/\sqrt{2} & 1/\sqrt{2} & 1/\sqrt{2} \end{bmatrix} \begin{bmatrix} v_a \\ v_b \\ v_c \end{bmatrix} \qquad (3.47)$$

Finally, an instantaneous voltage vector of the VSI can be expressed as (3.48).

$$\vec{v} = K_\alpha\vec{n}_\alpha + K_\beta\vec{n}_\beta + K_0\vec{n}_0 \qquad (3.48)$$

An instantaneous voltage vector of a three-leg center-split VSI can be represented by (3.49), where K_α, K_β and K_0 are $\sqrt{2/3}v_{dc}S_\alpha/2, 1/\sqrt{2}v_{dc}S_\beta/2$ and $1/\sqrt{3}v_{dc}S_0/2$ correspondingly. In (3.49) for simplicity, the upper voltage of the dc-link of the converter is assumed to be equal to the lower dc voltage of the VSI, $v_{dcU} = v_{dcL} = V_{dc}/2$.

$$\vec{v}_{center-split} = \frac{v_{dc}}{2}\left[\sqrt{\frac{2}{3}}S_\alpha \cdot \vec{n}_\alpha + \frac{1}{\sqrt{2}}S_\beta \cdot \vec{n}_\beta + \frac{1}{\sqrt{3}}S_0 \cdot \vec{n}_0\right] \qquad (3.49)$$

where $S_\alpha = S_a - \frac{1}{2}S_b - \frac{1}{2}S_c, S_\beta = S_b - S_c$ and $S_0 = S_a + S_b + S_c$.

3.3.2.1 Two-Level Three-Leg Center-Split VSI

For a two-level three-leg center-split VSI, there are totally eight output voltage vectors. The parameters of 3D voltage vectors in the a-b-c coordinates and in the α-β-0 coordinates are given in Table 3.4. The corresponding space vector allocation in 3D aspect is shown in Fig. 3.29. The vectors $\{\vec{V}_2, \vec{V}_4, \vec{V}_6\}$ and $\{\vec{V}_1, \vec{V}_3, \vec{V}_5\}$ lie on different horizontal levels and the zero vectors $\{\vec{V}_{0p}, \vec{V}_{0n}\}$ are the directional vectors pointing in positive and negative zero-axis respectively. The projection of the space vector allocation on the α-β plane is just the 2D space vector allocations, as shown in Fig. 3.30.

Based on the defined parameters in Cylindrical or Spherical Coordinates as given in (2.65), (2.66), (2.67) and (2.68), when $V_{dc} = 2$ is assumed, Table 3.5 summarizes two-level three-leg center-split VSI voltage vectors based on its cylindrical and spherical coordinate parameters. The mark, *, means "undefined" in the Table.

Table 3.4 3D voltage vectors of the two-level center-split VSI

Vectors	S_a	S_b	S_c	V_a	V_b	V_c	S_α	S_β	S_0	V_α	V_β	V_0
$\vec{V}_1$	1	−1	−1	$\frac{v_{dc}}{2}$	$-\frac{v_{dc}}{2}$	$-\frac{v_{dc}}{2}$	2	0	−1	$\sqrt{\frac{2}{3}}v_{dc}$	0	$\frac{-1}{2\sqrt{3}}V_{dc}$
$\vec{V}_2$	1	1	−1	$\frac{v_{dc}}{2}$	$\frac{v_{dc}}{2}$	$-\frac{v_{dc}}{2}$	1	2	1	$\sqrt{\frac{1}{6}}v_{dc}$	$\frac{1}{\sqrt{2}}v_{dc}$	$\frac{1}{2\sqrt{3}}V_{dc}$
$\vec{V}_3$	−1	1	−1	$-\frac{v_{dc}}{2}$	$\frac{v_{dc}}{2}$	$-\frac{v_{dc}}{2}$	−1	2	−1	$-\sqrt{\frac{1}{6}}v_{dc}$	$\frac{1}{\sqrt{2}}v_{dc}$	$\frac{-1}{2\sqrt{3}}V_{dc}$
$\vec{V}_4$	−1	1	1	$-\frac{v_{dc}}{2}$	$\frac{v_{dc}}{2}$	$\frac{v_{dc}}{2}$	−2	0	1	$-\sqrt{\frac{2}{3}}v_{dc}$	0	$\frac{1}{2\sqrt{3}}V_{dc}$
$\vec{V}_5$	−1	−1	1	$-\frac{v_{dc}}{2}$	$-\frac{v_{dc}}{2}$	$\frac{v_{dc}}{2}$	−1	−2	−1	$-\sqrt{\frac{1}{6}}v_{dc}$	$\frac{-1}{\sqrt{2}}v_{dc}$	$\frac{-1}{2\sqrt{3}}V_{dc}$
$\vec{V}_6$	1	−1	1	$\frac{v_{dc}}{2}$	$-\frac{v_{dc}}{2}$	$\frac{v_{dc}}{2}$	1	−2	1	$\sqrt{\frac{1}{6}}v_{dc}$	$\frac{-1}{\sqrt{2}}v_{dc}$	$\frac{1}{2\sqrt{3}}V_{dc}$
$\vec{V}_{00p}$	1	1	1	$\frac{v_{dc}}{2}$	$\frac{v_{dc}}{2}$	$\frac{v_{dc}}{2}$	0	0	3	0	0	$\frac{\sqrt{3}}{2}V_{dc}$
$\vec{V}_{00n}$	−1	−1	−1	$-\frac{v_{dc}}{2}$	$-\frac{v_{dc}}{2}$	$-\frac{v_{dc}}{2}$	0	0	−3	0	0	$\frac{-\sqrt{3}}{2}V_{dc}$

Fig. 3.29 3D space vector allocation of the two-level center-split VSI

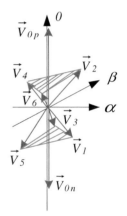

Fig. 3.30 Voltage space vector's allocation in the α-β frame

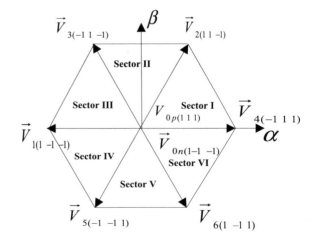

Table 3.5 Summarizes parameters of two-level voltage vectors for a two-level three-leg center-split converter

	S_a	S_b	S_c	S_α	S_β	S_0	r	θ	Z_0	ρ	ϕ
$\vec{V}_1$	1	−1	−1	2	0	−1	1.633	0°	−0.577	$\sqrt{3}$	109.46°
$\vec{V}_2$	1	1	−1	1	2	1	1.633	60°	0.577	$\sqrt{3}$	70.54°
$\vec{V}_3$	−1	1	−1	−1	2	−1	1.633	120°	−0.577	$\sqrt{3}$	109.46°
$\vec{V}_4$	−1	1	1	−2	0	1	1.633	180°	0.577	$\sqrt{3}$	70.54°
$\vec{V}_5$	−1	−1	1	−1	−2	−1	1.633	240°	−0.577	$\sqrt{3}$	109.46°
$\vec{V}_6$	1	−1	1	1	−2	1	1.633	300°	0.577	$\sqrt{3}$	70.54°
$\vec{V}_{0p}$	1	1	1	0	0	3	0	*	$\sqrt{3}$	$\sqrt{3}$	0°
$\vec{V}_{0n}$	−1	−1	−1	0	0	−3	0	*	$-\sqrt{3}$	$\sqrt{3}$	180°

The spherical and cylindrical coordinate parameters are given in Table when $V_{dc} = 2$

3.3.2.2   Three-Level Three-Leg Center-Split VSI

According to previous discussions in Chap. 2, there are $3^3 = 27$ space vectors for a three-level three-leg NPC inverter. If all these vectors are located in a two-dimensional plane, there are total 19 different vectors, since some of the vectors overlap on the α-β plane. The space vector allocation on the two-dimensional plane is shown in Fig. 3.31.

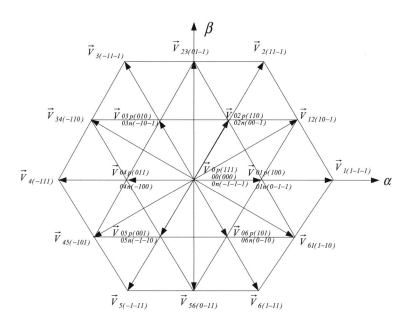

Fig. 3.31 Three-level voltage vectors allocation

According to the length of the space vectors, all these 27 vectors are classified to four groups. When $V_{dc} = 2$ is assumed, the detailed lists of the vectors in each group are provided in the Table 3.6, 3.7, 3.8 and 3.9.

- Zero-vector $\{\vec{V}_{00p}, \vec{V}_{000}, \vec{V}_{00n}\}$
- Small-vector $\{\vec{V}_{01p}, \vec{V}_{01n}, \vec{V}_{02p}, \vec{V}_{02n}, \vec{V}_{03p}, \vec{V}_{03n}, \vec{V}_{04p}, \vec{V}_{04n}, \vec{V}_{05p}\vec{V}_{05n}, \vec{V}_{06p}, \vec{V}_{06n}\}$
- Medium-vector $\{\vec{V}_{12}, \vec{V}_{23}, \vec{V}_{34}, \vec{V}_{45}, \vec{V}_{56}, \vec{V}_{61}\}$
- Large-vector $\{\vec{V}_1, \vec{V}_2, \vec{V}_3, \vec{V}_4, \vec{V}_5, \vec{V}_6\}$

If the system has accessible neutral wire, a zero-sequence current component can exist. It is desired that the load current zero-sequence component can be compensated by the active power filter or by the unbalance current compensator. In these cases, the zero-sequence current component, as well as the other components, must be controlled. The zero-sequence output of the voltage vectors are also provided in Tables 3.6, 3.7, 3.8 and 3.9. Figure 3.32 shows the voltage space vector in the α-β-0 frame.

Table 3.6 Zero vectors

	S_a	S_b	S_c	S_α	S_β	S_0	r	θ	Z_0	ρ	ϕ
$\vec{V}_{000}$	0	0	0	0	0	0	0	*	0	0	*
$\vec{V}_{00p}$	1	1	1	0	0	3	0	*	$\sqrt{3}$	$\sqrt{3}$	0°
$\vec{V}_{00n}$	−1	−1	−1	0	0	−3	0	*	$-\sqrt{3}$	$\sqrt{3}$	180°

Remark: "*" means "unavailable"

Table 3.7 Small vectors

	S_a	S_b	S_c	S_α	S_β	S_0	r	θ	Z_0	ρ	ϕ
$\vec{V}_{01p}$	1	0	0	1	0	1	0.816	0°	0.577	1	54.76°
$\vec{V}_{01n}$	0	−1	−1	1	0	−2	0.816	0°	−1.155	$\sqrt{2}$	144.76°
$\vec{V}_{02p}$	1	1	0	0.5	1	2	0.816	60°	1.155	$\sqrt{2}$	35.24°
$\vec{V}_{02n}$	0	0	−1	0.5	1	−1	0.816	60°	−0.577	1	125.24°
$\vec{V}_{03p}$	0	1	0	−0.5	1	1	0.816	120°	0.577	1	54.76°
$\vec{V}_{03n}$	−1	0	−1	−0.5	1	−2	0.816	120°	−1.155	$\sqrt{2}$	144.76°
$\vec{V}_{04p}$	0	1	1	−1	0	2	0.816	180°	1.155	$\sqrt{2}$	35.24°
$\vec{V}_{04n}$	−1	0	0	−1	0	−1	0.816	180°	−0.577	1	125.24°
$\vec{V}_{05p}$	0	0	1	−0.5	−1	1	0.816	240°	0.577	1	54.76°
$\vec{V}_{05n}$	−1	−1	0	−0.5	−1	−2	0.816	240°	−1.155	$\sqrt{2}$	144.76°
$\vec{V}_{06p}$	1	0	1	0.5	−1	2	0.816	300°	1.155	$\sqrt{2}$	35.24°
$\vec{V}_{06n}$	0	−1	0	0.5	−1	−1	0.816	300°	−0.577	1	125.24°

Table 3.8 Medium vectors

	S_a	S_b	S_c	S_α	S_β	S_0	r	θ	Z_o	ρ	ϕ
$\vec{V}_{12}$	1	0	−1	1.5	1	0	$\sqrt{2}$	30°	0	$\sqrt{2}$	90°
$\vec{V}_{23}$	0	1	−1	0	2	0	$\sqrt{2}$	90°	0	$\sqrt{2}$	90°
$\vec{V}_{34}$	−1	1	0	−1.5	1	0	$\sqrt{2}$	150°	0	$\sqrt{2}$	90°
$\vec{V}_{45}$	−1	0	1	−1.5	−1	0	$\sqrt{2}$	210°	0	$\sqrt{2}$	90°
$\vec{V}_{56}$	0	−1	1	0	−2	0	$\sqrt{2}$	270°	0	$\sqrt{2}$	90°
$\vec{V}_{61}$	1	−1	0	1.5	−1	0	$\sqrt{2}$	330°	0	$\sqrt{2}$	90°

Table 3.9 Large vectors

	S_a	S_b	S_c	S_α	S_β	S_0	r	θ	Z_0	ρ	ϕ
$\vec{V}_1$	1	−1	−1	2	0	−1	1.633	0°	−0.577	$\sqrt{3}$	109.46°
$\vec{V}_2$	1	1	−1	1	2	1	1.633	60°	0.577	$\sqrt{3}$	70.54°
$\vec{V}_3$	−1	1	−1	−1	2	−1	1.633	120°	−0.577	$\sqrt{3}$	109.46°
$\vec{V}_4$	−1	1	1	−2	0	1	1.633	180°	0.577	$\sqrt{3}$	70.54°
$\vec{V}_5$	−1	−1	1	−1	−2	−1	1.633	240°	−0.577	$\sqrt{3}$	109.46°
$\vec{V}_6$	1	−1	1	1	−2	1	1.633	300°	0.577	$\sqrt{3}$	70.54°

Fig. 3.32 Space vector allocation in the α-β-0 frame

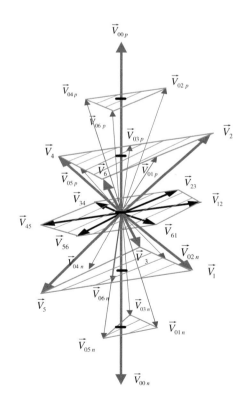

According to the values of the zero-sequence component of each vector, the total of 27 vectors can be classified to three groups.

- $Z_0 > 0$: $\vec{V}_{00p}$, $\left\{\vec{V}_{02p}, \vec{V}_{04p}, \vec{V}_{06p}\right\}$ and $\left\{\vec{V}_{01p}, \vec{V}_2, \vec{V}_{03p}, \vec{V}_4, \vec{V}_{05p}, \vec{V}_6\right\}$
- $Z_0 = 0$: $\vec{V}_{000}$ and $\left\{\vec{V}_{12}, \vec{V}_{23}, \vec{V}_{34}, \vec{V}_{45}, \vec{V}_{56}, \vec{V}_{61}\right\}$
- $Z_0 < 0$: $\vec{V}_{00n}$, $\left\{\vec{V}_{01n}, \vec{V}_{03n}, \vec{V}_{05n}\right\}$ and $\left\{\vec{V}_1, \vec{V}_{02n}, \vec{V}_3, \vec{V}_{04n}, \vec{V}_5, \vec{V}_{06n}\right\}$

There are seven voltage levels or units in zero-axis, which are $\{\sqrt{3},\ 1.155,\ 0.577,\ 0,\ -0.577,\ -1.155,\ -\sqrt{3}\}$ when $V_{dc} = 2$. All the medium vectors, $\left\{\vec{V}_{12}, \vec{V}_{23}, \vec{V}_{34}, \vec{V}_{45}, \vec{V}_{56}, \vec{V}_{61}\right\}$, locate on the horizontal plane of $Z_0 = 0$, form a hexagon, just like the space vector allocation of a two-level VSI. The current injecting to the system neutral wire is zero when the output vector of the three-level NPC inverter locates on this plane. As a result, the neutral current is not compensated.

When $Z_0 > 0$, there are three levels, as shown in Fig. 3.32. The vector $\vec{V}_{00p}$ has the highest zero-sequence output in which the outputs on the α-axis and β-axis are zero. For the other two levels, vectors $\left\{\vec{V}_{02p}, \vec{V}_{04p}, \vec{V}_{06p}\right\}$ and $\left\{\vec{V}_{01p}, \vec{V}_2, \vec{V}_{03p}, \vec{V}_4, \vec{V}_{05p}, \vec{V}_6\right\}$ can be utilized respectively. They are able to provide the zero-sequence output and provide compensation on the α-axis and β-axis simultaneously.

When $Z_0 < 0$, there are also three levels and the space vector allocation just symmetrical to the vectors when $Z_0 > 0$. In addition, there is one vector $\vec{V}_{000}$, which has zero output on the α-axis, β-axis and zero-axis.

3.3.3 Three-Dimensional Sign-Cubical Hysteresis PWM

3.3.3.1 Basic Control Strategy

In this section, a control strategy for the three-level VSI in a three-phase four-wire system is described by the Sign Cubical Hysteresis Current Controller or the Sign Rectangular Bar Hysteresis Current Controller [26]. The basic concept of this control strategy is explained in Fig. 3.33. The hysteresis limits of $\Delta\alpha$, $\Delta\beta$ and $\Delta 0$ can be equal to each other ($\Delta\alpha = \Delta\beta = \Delta 0$) so as to make use of the cubical control technique, however, they may not be equal to each other so that rectangular bar control hysteresis strategy is employed. The injected current of the three-level VSI is detected and transformed from a-b-c frame into α-β-0 frame.

The difference between the reference tracking current ($\vec{i}^*_{c\alpha\beta 0}$) and the actual compensating current ($\vec{i}_{c\alpha\beta 0}$) between the VSI and the point of common coupling terminal will be the control signal ($\Delta\vec{i}_{c\alpha\beta 0}$) for the controller to control the action of the VSI.

Fig. 3.33 Concept of sign cubical hysteresis control strategy

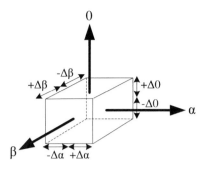

$$\Delta \vec{i}_{c\alpha\beta0} = \vec{i}^{*}_{c\alpha\beta0} - \vec{i}_{c\alpha\beta0} \tag{3.50}$$

There are three voltage levels $\{1, 0, -1\}$ in three-level VSI so that the sign of triggering pulses is an important parameter in tracking the reference current so as to choose the correct vectors.

When the reference compensating signal i^{*}_{cx} and the actual compensating current i_{cx} are compared, three directions can be defined.

- If $i^{*}_{cx} > i_{cx} + \Delta e (x = \alpha, \beta, 0)$ the compensating direction is defined as $+1$;
- If $i_{cx} - \Delta e < i^{*}_{cx} < i_{cx} + \Delta e$ ($x = \alpha, \beta, 0$), the compensating direction is defined as 0;
- If $i^{*}_{cx} < i_{cx} - \Delta e$ ($x = \alpha, \beta, 0$), the compensating direction is defined as -1.

When the difference between the reference signal and actual input signal is larger than the hysteresis band value, the output voltage of this leg is triggered to a positive or, vice versa, to a negative level. However, when the difference is less than the hysteresis band value, there will be the zero level. According to the direction of the current compensating requirement, the output voltage vectors of the three-level VSI can be selected among the 27 available output voltage vectors.

3.3.3.2 Switching Tables and Selection

The compensating direction of each output vector of the three-level VSI is summarized in the Table 3.10. In order to select the proper output vector of the three-level VSI to achieve the required compensation target, more detailed information about the space vector is provided hereinafter.

(1) Vector-Pairs

In Table 3.10 that there are six pairs of vectors, of which two vectors have the same compensation direction. The six pairs are listed in Table 3.11.

When the required compensating direction is the same as the direction of the vectors of one pair, one vector has to be chosen from these two vectors. Although the two vectors have the same directions, their amplitudes in each

Table 3.10 Compensation directions of the three-dimensional space vectors

		S_α	S_β	S_0			S_α	S_β	S_0
Large vectors	$\vec{V}_1(1,-1,-1)$	+	0	-	Small vectors	$\vec{V}_{01p}(1,0,0)$	+	0	+
	$\vec{V}_2(1,1,-1)$	+	+	+		$\vec{V}_{01n}(0,-1,-1)$	+	0	-
	$\vec{V}_3(-1,1,-1)$	-	+	-		$\vec{V}_{02p}(1,1,0)$	+	+	+
	$\vec{V}_4(-1,1,1)$	-	0	+		$\vec{V}_{02n}(0,0,-1)$	+	+	-
	$\vec{V}_5(-1,-1,1)$	-	-	-		$\vec{V}_{03p}(0,1,0)$	-	+	+
	$\vec{V}_6(1,-1,1)$	+	-	+		$\vec{V}_{03n}(-1,0,-1)$	-	+	-
Medium vectors	$\vec{V}_{12}(1,0,-1)$	+	+	0		$\vec{V}_{04p}(0,1,1)$	-	0	+
	$\vec{V}_{23}(0,1,-1)$	0	+	0		$\vec{V}_{04n}(-1,0,0)$	-	0	-
	$\vec{V}_{34}(-1,1,0)$	-	+	0		$\vec{V}_{05p}(0,0,1)$	-	-	+
	$\vec{V}_{45}(-1,0,1)$	-	-	0		$\vec{V}_{05n}(-1,-1,0)$	-	-	-
	$\vec{V}_{56}(0,-1,1)$	0	-	0		$\vec{V}_{06p}(1,0,1)$	+	-	+
	$\vec{V}_{61}(1,-1,0)$	+	-	0		$\vec{V}_{06n}(0,-1,0)$	+	-	-
Zero vectors	$\vec{V}_{00p}(1,1,1)$	0	0	+					
	$\vec{V}_{00n}(-1,-1,-1)$	0	0	-					
	$\vec{V}_{000}(0,0,0)$	0	0	0					

Table 3.11 Vector pair in the same direction

	S_α	S_β	S_0	Large vectors	Small vectors
1	+	0	-	$\vec{V}_1$	$\vec{V}_{01n}$
2	+	+	+	$\vec{V}_2$	$\vec{V}_{02p}$
3	-	+	-	$\vec{V}_3$	$\vec{V}_{03n}$
4	-	0	+	$\vec{V}_4$	$\vec{V}_{04p}$
5	-	-	-	$\vec{V}_5$	$\vec{V}_{05n}$
6	+	-	+	$\vec{V}_6$	$\vec{V}_{06p}$

direction are different. The output of the pair vectors on the α–β plane and in the zero-axis can be found in the Tables 3.7 and 3.9. The amplitude comparison of vector pair is listed in Table 3.12, in which it is assumed $V_{dc} = 2$, $V_{dcu} = V_{dcL} = 1$.

It shows that the vector from the large vector group contributes more in the α–β frame, but provides less output in the zero-axis. The vector from the small vector group provides higher zero-axis compensation capability, but its output in the α–β frame is smaller. When one vector is chosen from these

Table 3.12 Amplitude comparison of vector-pair

	$V_{\alpha\beta}$	V_0
$\vec{V}_1, \vec{V}_2, \vec{V}_3, \vec{V}_4, \vec{V}_5, \vec{V}_6$	1.633	0.577
$\vec{V}_{01n}, \vec{V}_{02p}, \vec{V}_{03n}, \vec{V}_{04p}, \vec{V}_{05n}, \vec{V}_{06p}$	0.816	1.154

pair-vectors, the error amplitude of $\sqrt{\left(\Delta i_\alpha\right)^2 + \left(\Delta i_\beta\right)^2}$ is compared with Δi_0, the largest one is chosen so that the vector can be determined from those pair-vectors to reduce error.

(2) Special Vectors

The number of available output vectors of a three-level inverter is limited. For the required compensation direction in Table 3.13, no vector can satisfy the requirement of all the three directions. The vector that is used can only satisfy the requirement in one or two directions. As a result, errors are introduced in the remained direction. In Table 3.13, four possible selective vectors are listed for each case. For example, when the required compensation direction is $S_\alpha > 0$, $S_\beta = 0$ and $S_0 = 0$, the vector can be selected from, $\vec{V}_1, \vec{V}_{12}, \vec{V}_{01p}$ or $\vec{V}_{01n}$. The output of those four vectors can be found in Table 3.6, 3.7, 3.8, 3.9, which   are

$\vec{V}_1 = \left\{S_\alpha, S_\beta, S_0\right\} = \{\,2 \quad 0 \quad -1\,\}$, $\vec{V}_{12} = \left\{S_\alpha, S_\beta, S_0\right\} = \{\,1.5 \quad 1 \quad 0\,\}$, $\vec{V}_{01P} = \left\{S_\alpha, S_\beta, S_0\right\} = \{\,1 \quad 0 \quad 1\,\}$ and $\vec{V}_{01n} = \left\{S_\alpha, S_\beta, S_0\right\} = \{\,1 \quad 0 \quad -2\,\}$.

Their outputs in three directions are different. $\vec{V}_1$ provides the highest improvement in α-axis and introduces error in the zero-axis. Based on the regulation that the compensation on the α-β plane is put on a higher priority, $\vec{V}_1$ is selected among those vectors. In other cases shown in Table 3.13, the vector from the first column is selected.

(3) Final Table:

There are totally 27 possible combinations in three-level VSI. Table 3.14 lists the final switching table in the sign-cubic hysteresis PWM. The control scheme is shown in Fig. 3.34. The sign of the S_α, S_β, S_0 are important parameters to determine the voltage vector. When the sign is positive or negative, it means that the signal is larger than the hysteresis limit. When it is zero, the error signal is within the hysteresis limit. According to Table 3.14, the switching sequence for the three-level VSI can be determined uniquely.

3.3.3.3 Simulation Results

The simulation is performed by MATLAB/Simulink. The system configuration is shown in Fig. 3.27, which is a three-phase four-wire active power filter using a

Table 3.13 Special vectors

	S_α	S_β	S_0	Possible selection vectors			
1	+	0	0	$\vec{V}_1$	$\vec{V}_{12}$	$\vec{V}_{01p}$	$\vec{V}_{01n}$
2	−	0	0	$\vec{V}_4$	$\vec{V}_{45}$	$\vec{V}_{04p}$	$\vec{V}_{04n}$
3	0	+	+	$\vec{V}_{23}$	$\vec{V}_{00p}$	$\vec{V}_{02p}$	$\vec{V}_{03p}$
4	0	+	−	$\vec{V}_{23}$	$\vec{V}_{00n}$	$\vec{V}_{02n}$	$\vec{V}_{03n}$
5	0	−	−	$\vec{V}_{56}$	$\vec{V}_{00n}$	$\vec{V}_{06n}$	$\vec{V}_{05n}$
6	0	−	+	$\vec{V}_{56}$	$\vec{V}_{00p}$	$\vec{V}_{06p}$	$\vec{V}_{05p}$

Table 3.14 Final table

	S_α	S_β	S_0	S_a	S_b	S_c		S_α	S_β	S_0	S_a	S_b	S_c	S_a	S_b	S_c
1	+	+	−	0	0	−1	15	0	0	0	0	0	0			
2	+	−	−	0	−1	0	16	+	0	0	1	−1	−1			
3	−	+	+	0	1	0	17	−	0	0	−1	1	1			
4	−	−	+	0	0	1	18	0	+	+	0	1	−1			
5	+	+	0	1	0	−1	19	0	+	−	0	1	−			
6	+	−	0	1	−1	0	20	0	−	−	0	−1	1			
7	−	+	0	−1	1	0	21	0	−	+	0	−1	1			
8	−	−	0	−1	0	1					Larger $V_{\alpha\beta}$			Larger V_0		
9	0	+	0	0	1	−1	22	+	+	+	1	1	−1	1	1	0
10	0	−	0	0	−1	1	23	+	−	+	1	−1	1	1	0	1
11	+	0	+	1	0	0	24	−	+	−	−1	1	−1	−1	0	−1
12	−	0	−	−1	0	0	25	−	−	−	−1	−1	1	−1	−1	0
13	0	0	+	1	1	1	26	+	0	−	1	−1	−1	0	−1	−1
14	0	−	−	−1	−1	−1	27	−	0	+	−1	1	1	0	1	1

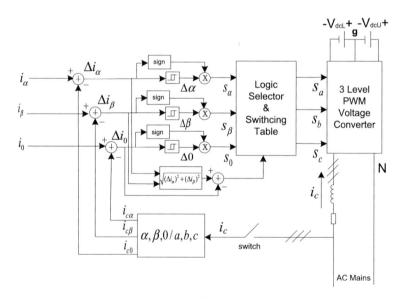

Fig. 3.34 Control scheme of 3D sign-cubic hysteresis PWM

three-level VSI. The sampling frequency to determine the output vector of the three-level VSI is 20 kHz. This is also the upper-limit of the switching frequency for each power switch. The actual switching frequency is lower than this value since the output vector may not be changed in every period. The selected hysteresis band affects the actual switching frequency. Larger hysteresis band leads to lower switching frequency and vice versa.

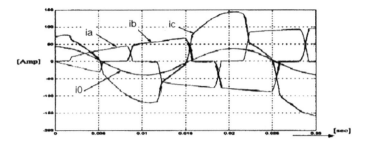

Fig. 3.35   Load currents without compensation

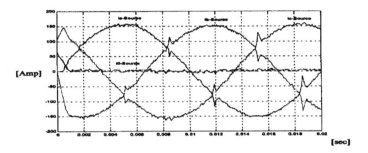

Fig. 3.36   Current waveforms after compensation

Figure 3.35 shows the current waveforms before compensation. The current after compensation is shown in Fig. 3.36. Results indicate that the 3D sign-cubic hysteresis PWM can control the three-level inverter to achieve a satisfying result in a three-phase four-wire active power filter.

3.3.4   Three-Dimensional Cylindrical Coordinate PWM

The main drawbacks of the sign cubical hysteresis control strategy are as follows:

- The special vectors divert one improvement in one direction and give dedicated error in another direction.
- Switching frequency is relatively high and random.
- The compensated results may be unstable due to the overcompensation or under-compensation when the switching frequency is fixed and the results are easy to be affected by the system's parameters.

In this section, the cylindrical coordinate control 3DPWM is adopted for reducing the switching frequency and eliminating special vector effects defined in the last section. In this PWM method, the reference voltage vector is used to determine the output vector of the VSI instead of the current reference used in the 3D sign-cubic hysteresis PWM. The voltage reference can be calculated according to the required current reference as discussed in Sect. 3.2.2. The 3D space vectors in the cylindrical coordinates are analyzed first before the cylindrical coordinate control of 3DPWM [10] is discussed.

3.3.4.1   Cylindrical Coordinate Voltage Vectors

In 3D aspect, voltage vectors can be expressed in rectangular, cylindrical or spherical coordinates. In the rectangular coordinate, $\left\{ \vec{n}_\alpha, \vec{n}_\beta, \vec{n}_0 \right\}$ form a basis and they are orthogonal to each other such that $\vec{n}_0 \cdot \vec{n}_\alpha = \vec{n}_0 \cdot \vec{n}_\beta = \vec{n}_\alpha \cdot \vec{n}_\beta = 0$. The instantaneous voltage vector can also be expressed as

$$\vec{v} = v_\alpha \cdot \vec{n}_\alpha + v_\beta \cdot \vec{n}_\beta + v_0 \cdot \vec{n}_0 \tag{3.51}$$

where $v_\alpha = V_{dc}/2 \cdot Z_\alpha$, $v_\beta = V_{dc}/2 \cdot Z_\beta$ and $v_0 = V_{dc}/2 \cdot Z_0$.

The parameters describing the rectangular coordinate are Z_α, Z_β, Z_0 and the amplitude ρ is expressed as (3.52).

$$\rho = \sqrt{Z_\alpha^2 + Z_\beta^2 + Z_0^2} \tag{3.52}$$

where $Z_\alpha = \sqrt{\frac{2}{3}} S_\alpha$, $Z_\beta = \frac{1}{\sqrt{2}} S_\beta$, $Z_0 = \frac{1}{\sqrt{3}} S_0$ and

$$S_\alpha = S_a - \frac{1}{2} S_b - \frac{1}{2} S_c$$
$$S_\beta = S_b - S_c$$
$$S_0 = S_a + S_b + S_c$$

In cylindrical coordinate, each point in the 3D space is denoted by three parameters, $\{r, \theta, z_0\}$, as shown in Fig. 3.37. The corresponding definitions for each parameter are given in (3.53) to (3.55). The values of these parameters for each vector of a three-level VSI are also listed in Table 3.6, 3.7, 3.8, 3.9.

$$r = \sqrt{Z_\alpha^2 + Z_\beta^2} \tag{3.53}$$

$$\theta = \tan^{-1} \left(\frac{Z_\beta}{Z_\alpha} \right) \tag{3.54}$$

$$Z_0 = \frac{1}{\sqrt{3}} S_0 \tag{3.55}$$

Fig. 3.37 Cylindrical coordinate

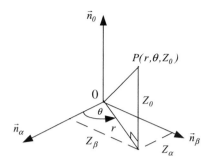

3.3.4.2 Switching Tables for the Cylindrical Coordinate PWM

After the reference voltage vector and the space vectors of the three-level VSI are expressed in the cylindrical coordinates, switching tables are designed for selecting the proper output vector for the three-level VSI so that the required output voltage of the VSI can be generated. (3.56) and (3.57) are the polar voltage value and polar angle respectively.

$$v_r = \sqrt{Z_\alpha^2 + Z_\beta^2} \tag{3.56}$$

$$\theta_r = \tan^{-1}\left(\frac{Z_\beta}{Z_\alpha}\right) \tag{3.57}$$

The switching tables are designed according to the nearest voltage vector in four different cases. It is assumed that the voltage is normalized by $v_{dcU} = v_{dcL} = v_{dc/2} = 1$. The zero sequence of the reference voltage vector v_0 equals to the Z_0 in the cylindrical coordinates. There are four values of v_r for the output voltage vector $r \in \{0.816, 1.414, 1.633\}$. As a result, four switching tables are classified according to the polar voltage value v_r.

Case 1: Very Small Polar Voltage, v_r

When the normalized polar voltage is within $v_r \in \{0, 0.408\}$, it is not necessary to compensate the α-axis and β-axis components. Only zero voltage compensation is considered and one zero voltage vector is activated. The zero voltage vectors include $\vec{v}_{00p}$, $\vec{v}_{000}$ and $\vec{v}_{00n}$. If $v_0 \in \{-0.866, 0.866\}$, $\vec{v}_{000}$ is chosen. When $v_0 < -0.866$, $\vec{v}_{00n}$ is used. When $v_0 > 0.866$, $\vec{v}_{00p}$ is activated. When the required polar voltage is within $\{0, 0.408\}$ and no matter what the polar angle θ_r is, the output voltage vector is selected according to Table 3.15.

Table 3.15 Very small v_r—zero vectors

$|v_r| \in \{0, 0.408\}$	$v_0 < -0.866$	$v_0 \in \{-0.866, 0.866\}$	$v_0 > 0.866$
Any θ	$\vec{V}_{00n} = \begin{bmatrix} -1 & -1 & -1 \end{bmatrix}$	$\vec{V}_{000} = \begin{bmatrix} 0 & 0 & 0 \end{bmatrix}$	$\vec{V}_{00p} = \begin{bmatrix} 1 & 1 & 1 \end{bmatrix}$

Case 2: Small Polar Voltage, v_r

When $v_r \in \{0.408, 1.115\}$, the small polar voltage is defined, in this case, polar angle θ_r and zero voltage v_0 also affect the selection of the output voltage vector of the VSI. Referring to the Table 3.7 and Fig. 3.32, the vectors $\vec{v}_{01p}, \vec{v}_{02p}, \vec{v}_{03p}, \vec{v}_{04p}, \vec{v}_{05p}$ and $\vec{v}_{06p}$ are allocated in the positive zero axis although there are two different zero levels. The zero-sequence output of the vectors $\vec{v}_{02p}, \vec{v}_{04p}$ and $\vec{v}_{06p}$ are higher than that of the vectors $\vec{v}_{01p}, \vec{v}_{03p}$ and $\vec{v}_{05p}$. On the other hand, there are vectors $\vec{v}_{01n}, \vec{v}_{02n}, \vec{v}_{03n}, \vec{v}_{04n}, \vec{v}_{05n}$ and $\vec{v}_{06n}$ allocated in the negative zero axis. In this small polar voltage case, the nearest polar voltage vector is chosen so that when the polar angle of the reference voltage is within $\theta_r \in \{330°, 30°\}$, one output vector is selected from $\vec{v}_{01p}$ and $\vec{v}_{01n}$ according to the zero-sequence direction. If the zero voltage v_0 is positive, $\vec{v}_{01p}$ can be activated. Otherwise, $\vec{v}_{01n}$ is chosen. According to the above consideration, Table 3.16 is defined.

Case 3: Medium Polar Voltage v_r

When $v_r \in \{1.115, 1.524\}$, the required polar voltage is defined as medium polar voltage. From Fig. 3.32, it is noticed that the medium voltage vectors are allocated on the zero voltage level. The medium vectors $\{\vec{v}_{12}, \vec{v}_{23}, \vec{v}_{34}, \vec{v}_{45}, \vec{v}_{56}, \vec{v}_{61}\}$ are laid on the α-β plane. The medium vectors do not provide compensation capability on the zero-axis. The output voltage vector is chosen according to the polar angle when the zero-axis voltage locates in the range of $v_0 = \{-0.289, 0.289\}$. The switching table for the voltage vectors is given in Table 3.17. However, when the zero-axis voltage of the reference voltage vector is large, the zero voltage vectors are chosen instead. The zero voltage error is forced to be reduced first and reaches a small value, e.g. 0.289. Then, medium voltage vectors are activated. According to the above logic, Table 3.17 is defined for medium required voltage. Referring to above section, the Sign Cubical Hysteresis Control Strategy can reduce the error in one direction but it can increase or dedicate the error in another direction. There is no perfectly matched vector in all α, β and 0 frames in Sign Cubical Hysteresis Control. However, in this proposed strategy, the error can be reduced in one direction but it cannot inject the error in another.

Case 4: Large Polar Voltage, v_r

When $v_r > 1.524$, large voltage and small voltage vectors are chosen to approximate the reference voltage vector. Large voltage vectors are preferred since

Table 3.16 Smal v_r—small vectors

$|v_r| \in \{0.408, \quad 1.115\}$	$v_0 \geq 0$	$v_0 < 0$
$\theta \in \{330°, 30°\}$	$\vec{V}_{01p} = [1 \quad 0 \quad 0]$	$\vec{V}_{01n} = [0 \quad -1 \quad -1]$
$\theta \in \{30°, 90°\}$	$\vec{V}_{02p} = [1 \quad 1 \quad 0]$	$\vec{V}_{02n} = [0 \quad 0 \quad -1]$
$\theta \in \{90°, 150°\}$	$\vec{V}_{03p} = [0 \quad 1 \quad 0]$	$\vec{V}_{03n} = [-1 \quad 0 \quad -1]$
$\theta \in \{150°, 210°\}$	$\vec{V}_{04p} = [0 \quad 1 \quad 1]$	$\vec{V}_{04n} = [-1 \quad 0 \quad 0]$
$\theta \in \{210°, 270°\}$	$\vec{V}_{05p} = [0 \quad 0 \quad 1]$	$\vec{V}_{05n} = [-1 \quad -1 \quad 0]$
$\theta \in \{270°, 330°\}$	$\vec{V}_{06p} = [1 \quad 0 \quad 1]$	$\vec{V}_{06n} = [0 \quad -1 \quad 0]$

Table 3.17 Medium v_r—medium and zero vectors

$\|v_r\| \in \{1.115, \quad 1.524\}$	$v_0 < -0.289$	$-0.289 \leq v_0 \leq 0.289$	$v_0 > 0.289$
$\theta \in \{0°, 60°\}$	$\vec{V}_{00n} = [-1 \quad -1 \quad -1]$	$\vec{V}_{12} = [1 \quad 0 \quad -1]$	$\vec{V}_{00p} = [1 \quad 1 \quad 1]$
$\theta \in \{60°, 120°\}$	$\vec{V}_{00n} = [-1 \quad -1 \quad -1]$	$\vec{V}_{23} = [0 \quad 1 \quad -1]$	$\vec{V}_{00p} = [1 \quad 1 \quad 1]$
$\theta \in \{120°, 180°\}$	$\vec{V}_{00n} = [-1 \quad -1 \quad -1]$	$\vec{V}_{34} = [-1 \quad 1 \quad 0]$	$\vec{V}_{00p} = [1 \quad 1 \quad 1]$
$\theta \in \{180°, 240°\}$	$\vec{V}_{00n} = [-1 \quad -1 \quad -1]$	$\vec{V}_{45} = [-1 \quad 0 \quad 1]$	$\vec{V}_{00p} = [1 \quad 1 \quad 1]$
$\theta \in \{240°, 300°\}$	$\vec{V}_{00n} = [-1 \quad -1 \quad -1]$	$\vec{V}_{56} = [0 \quad -1 \quad 1]$	$\vec{V}_{00p} = [1 \quad 1 \quad 1]$
$\theta \in \{300°, 360°\}$	$\vec{V}_{00n} = [-1 \quad -1 \quad -1]$	$\vec{V}_{61} = [1 \quad -1 \quad 0]$	$\vec{V}_{00p} = [1 \quad 1 \quad 1]$

they have better compensation capability. However, there are limited numbers of large vectors that can be supplied in a three-level VSI. Small voltage vectors are chosen instead when the polar angle and zero-axis voltage locates in certain range. As a result, the nearest polar voltage vector is chosen in a three-level VSI case according to the reference voltage vector. The switching table is designed and is given in Table 3.18

From the above discussion, the polar voltage v_r, polar angle θ_r, and zero voltage v_0 are the parameters to design the activated vector. Those three parameters are dedicated to represent a 3D vector in cylindrical coordinate. As there are limited number of supplied voltage vectors by a three-level VSI, the nearest polar vector is chosen according to the polar voltage amplitude, polar angle and zero voltage value of the reference voltage vector. The block diagram for the 3D cylindrical coordinate PWM is shown in Fig. 3.38

3.3.4.3 Simulation Results

The simulation is performed by MATLAB/SIMULINK as the power quality compensator with a fixed switching frequency. A simulation test is performed by applying cylindrical coordinate control of a 3-D pulse width modulation in a three-phase four-wire active power filter, in which a three-level VSI is used. Figure 3.39 shows the nonlinear load current before the compensation. The current unbalance, harmonic and neutral currents needed to be compensated. Figure 3.40

Table 3.18 Large v_r—large and small vectors

$\|v_r\| > 1.524$	$v_0 \geq 0$	$v_0 < 0$
$\theta \in \{330°, 30°\}$	$\vec{V}_{01p} = [1 \quad 0 \quad 0]$	$\vec{V}_1 = [1 \quad -1 \quad -1]$
$\theta \in \{30°, 90°\}$	$\vec{V}_2 = [1 \quad 1 \quad -1]$	$\vec{V}_{02n} = [0 \quad 0 \quad -1]$
$\theta \in \{90°, 150°\}$	$\vec{V}_{03p} = [0 \quad 1 \quad 0]$	$\vec{V}_3 = [-1 \quad 1 \quad -1]$
$\theta \in \{150°, 210°\}$	$\vec{V}_4 = [-1 \quad 1 \quad 1]$	$\vec{V}_{04n} = [-1 \quad 0 \quad 0]$
$\theta \in \{210°, 270°\}$	$\vec{V}_{05p} = [0 \quad 0 \quad 1]$	$\vec{V}_5 = [-1 \quad -1 \quad 1]$
$\theta \in \{270°, 330°\}$	$\vec{V}_6 = [1 \quad -1 \quad 1]$	$\vec{V}_{06n} = [0 \quad -1 \quad 0]$

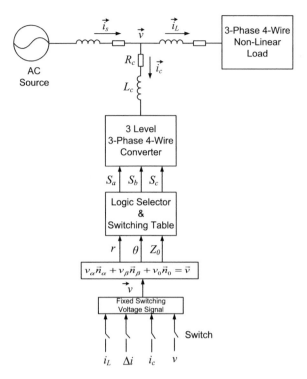

Fig. 3.38 Control block diagram of the cylindrical coordinate PWM

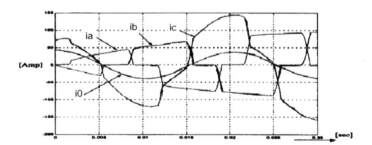

Fig. 3.39 Load current before compensation

shows the source currents after compensation. It is obvious that the 3D cylindrical coordinate PWM could control the APF to reduce the harmonic spectrum and to compensate the reactive and unbalance currents with eliminating neutral current as well.

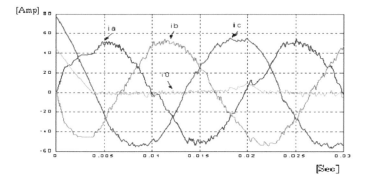

Fig. 3.40 Source current after compensation

3.3.4.4 Comparison of 3D Sign Cubical Hysteresis and Cylindrical Coordinate PWM

The performance of 3D sign cubical hysteresis control and cylindrical coordinate PWM are compared in this section. The performance indices are proposed in (3.58) to (3.61) in order to compare the performance of the above two 3D PWM methods. The indices J_α, J_β and J_0 are the average absolute error in the α, β and 0 frames of one period. The index, $J_{\alpha\beta0}$ is the sum of all absolute mean error. The performance is better when the value of the indices is smaller.

$$J_\alpha = \frac{1}{T} \int_0^T |\Delta i_\alpha| dt \tag{3.58}$$

$$J_\beta = \frac{1}{T} \int_0^T |\Delta i_\beta| dt \tag{3.59}$$

$$J_0 = \frac{1}{T} \int_0^T |\Delta i_0| dt \tag{3.60}$$

$$J_{\alpha\beta0} = \frac{1}{T} \int_0^T \left(|\Delta i_\alpha| + |\Delta i_\beta| + |\Delta i_0| \right) dt \tag{3.61}$$

Figures 3.41 and 3.42 show the waveforms of Δi_α, Δi_β and Δi_0 by 20 kHz sign cubical control and 10 kHz cylindrical coordinate PWM for compensation respectively. Table 3.19 summarizes the values obtained by the simulation according to the proposed performance indices and total harmonics distortion

Fig. 3.41 Variation of
performance indices for the
20 kHz 3D sign cubical
hysteresis PWM

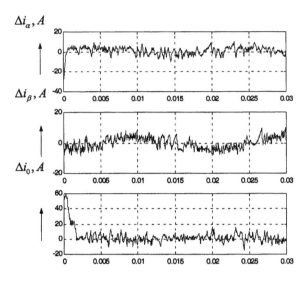

Fig. 3.42 Variation of
performance indices for the
10 kHz cylindrical coordinate
PWM

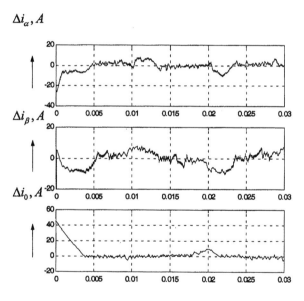

(THD) of compensated current waveforms in phase a, b and c. Comparing the THD values of the line currents, the performance of 10 kHz Cylindrical Coordinate Control is better than that of the 20 kHz Sign Cubical by about 37 % in average. On the other hand, there are respectively around 79.12 and 84.95 % improvement in the value of $J_{\alpha\beta0}$ in the Sign Cubical Control and the Cylindrical Coordinate Control. Checking the values of J_α, J_β and J_0, the compensated improvement is almost the same inα, β and 0 axes of the Sign Cubical. However, there is better improvement in the zero-axis by the Cylindrical Coordinate Control Strategy.

Table 3.19 Comparison of 3D PWM performance

Comparison of 3D PWM performance	Without compensation	3D sign cubical hysteresis PWM 20 kHz	3D cylindrical coordinate PWM 10 kHz
THD of phase a current (%)	30.47	11.82	7.17
THD of phase b current (%)	32.22	10.40	5.72
THD of phase c current (%)	41.47	8.49	6.26
J_α	15.59	3.06	2.25
J_β	12.13	3.09	3.39
J_0	18.67	3.50	1.34
$J_{\alpha\beta0}$	46.39	9.69	6.98

3.3.5 Three-Dimensional Space Vector Modulation

The basic Three-Dimensional Space Vector Modulation (3DSVM) for a two-level inverter is given in Section A and then the corresponding generalized SVM for a three-level inverter is discussed in Section B.

3.3.5.1 Basic 3D Two-Level Space Vector Modulation

The Space Vector Modulation is based on the volt-second approximation. However, in the 3DSVM of a two-level three-leg centre-split VSI, all the eight vectors contribute to the zero-sequence compensation, which is different from the conventional 2DSVM. In 3DSVM [27], the volt-second reference can be approximated by a sequence of volt-second states as given in (3.62) and (3.63).

$$\vec{V}_{ref}T_S = \vec{V}_x t_x + \vec{V}_y t_y + \vec{V}_0 t_0 + \vec{V}_{zero} t_{zero} \tag{3.62}$$

$$t_{zero} = T_S - t_x - t_y - t_0 \tag{3.63}$$

The vectors, $\vec{V}_x, \vec{V}_y$ and $\vec{V}_0$, are chosen as they are the neighboring vectors of reference vector on α-β plane. The neighboring vectors on the apex of the hexagon, as shown in Fig. 3.30, have the property that when the output of the inverter changes from one vector to another neighboring one, only the switching state of one leg needs to be changed accordingly. Based on this property, the optimum switching sequence scheme can be implemented and the output harmonics of the inverter can be reduced. However, the vectors, $\vec{V}_x$ and $\vec{V}_y$, contribute to zero sequence component as well as the α-β plane compensation to the reference vector. The vector $\vec{V}_0$ contributes to the zero sequence compensation only and can be chosen according to the required zero-axis or the neutral current compensating component. The product of $\vec{V}_{zero} t_{zero}$ should be equal to zero in the sense that t_{zero} is the redundant time in one compensation period. In the case of over-modulation, t_{zero} is equal to zero.

Actually, the reference vector $\vec{V}_{ref}$ can be described in three-dimensional aspect as shown in Fig. 3.43:

$$\vec{V}_{ref} = \vec{V}_\alpha \vec{n}_\alpha + \vec{V}_\beta \vec{n}_\beta + \vec{V}_0 \vec{n}_0 \tag{3.64}$$

The formula for calculating the switching time t_x, t_y and t_0 can be expressed in a general form (3.65). The matrix [Ag] can be expressed in six forms according to the sector location of reference vector $\vec{V}_{ref}$. After the sector location of $\vec{V}_{ref}$ is detected in Fig. 3.30, the corresponding section location value "g" can be determined.

$$[V_{ref}] = m[A_g][t_{xy0}]$$
$$g = \text{I, II, III, IV, V or VI.}$$

$$A_I = \begin{bmatrix} 1 & \cos 60° & 0 \\ 0 & \sin 60° & 0 \\ -C\cos k & C\cos k & C \end{bmatrix} A_{II} = \begin{bmatrix} \cos 60° & -\cos 60° & 0 \\ \cos 30° & \cos 30° & 0 \\ C\cos k & -C\cos k & C \end{bmatrix} A_{III} = \begin{bmatrix} -\cos 60° & -1 & 0 \\ \cos 30° & 0 & 0 \\ -C\cos k & C\cos k & C \end{bmatrix}$$

$$A_{IV} = \begin{bmatrix} -1 & -\cos 60° & 0 \\ 0 & -\cos 30° & 0 \\ C\cos k & -C\cos k & C \end{bmatrix} A_{II} = \begin{bmatrix} -\cos 60° & \cos 60° & 0 \\ -\cos 30° & -\cos 30° & 0 \\ -C\cos k & C\cos k & C \end{bmatrix} A_{III} = \begin{bmatrix} \cos 60° & 1 & 0 \\ -\cos 30° & 0 & 0 \\ C\cos k & -C\cos k & C \end{bmatrix}$$

$$\tag{3.65}$$

where $m = \frac{2\cdot\sqrt{\frac{2}{3}}\cdot V_{dc}}{T_s}$, $C = \frac{3}{2\cdot\sqrt{2}}$ and $k = 70.54°$.

In a conventional two-dimensional pulse width modulation technique, among the eight available voltage vectors, there are 6 directional vectors and 2 zero vectors, $\vec{V}_{op}$ and $\vec{V}_{on}$. Since no injecting action is performed by choosing $\vec{V}_{op}$ or $\vec{V}_{on}$, either of these two zero vectors can be substituted as a net zero vector, $\vec{V}_{zero} = 0$. And sometimes both of the two vectors are chosen according to the requirement of the sequence scheme. However, in 3D two-level center-split VSI system, there are 8 directional vectors corresponding to 8 available vectors. When referred to Fig. 3.29,

Fig. 3.43 The decomposition of reference voltage vector in 3D

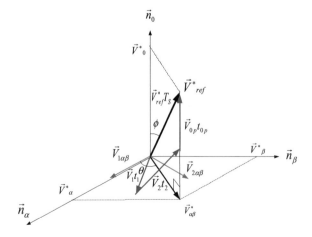

those vectors, $\vec{V}_{op}$ and $\vec{V}_{on}$, are not zero. In 3DPWM, these two vectors are dedicated as the zero-axis voltage components in the positive or negative direction and they are used to compensate neutral current in a three-phase four-wire system. The issue of $\vec{V}_{zero}t_{zero}$ can be solved by the exercise of the same amount of switching times, $t_{op} = t_{on}$, so that the approximated procedure for $\vec{V}_{zero}t_{zero} = 0$ can be obtained as:

$$\vec{V}_{op}t_{op} + \vec{V}_{on}t_{on} \approx \vec{V}_{zero}t_{zero} \approx 0 \qquad (3.66)$$

Furthermore, according to the compensation requirement, $\vec{V}_{on}$ is activated for negative zero-current injection into the neutral wire and $\vec{V}_{op}$ is activated for positive zero-current injection so that $t_{op} \neq t_{on}$ may occur for the neutral current compensation. In general, Eq. (3.62) should be replaced by (3.67).

$$\vec{V}_{ref}T_s = \vec{V}_x t_x + \vec{V}_y t_y + \vec{V}_{on}t_{on} + \vec{V}_{op}t_{op} \qquad (3.67)$$

The switching times are calculated under two conditions: Under Modulation and Over-Modulation.
Case 1: Under Modulation
After t_x, t_y, and t_0 are obtained from Eq. (3.65), t_{op} and t_{on} in Eq. (3.67) can be decided accordingly. The consideration of $\vec{V}_{zero}t_{zero}$ is taken as an actual null, i.e., there must be null output in all axes.
Positive Zero Vector is required in:

$$\begin{cases} t_{op} = t_0 + \frac{t_{zero}}{2} \\ t_{on} = \frac{t_{zero}}{2} \end{cases} \qquad (3.68)$$

Whereas, negative Zero Vector is required in:

$$\begin{cases} t_{op} = \frac{t_{zero}}{2} \\ t_{on} = t_0 + \frac{t_{zero}}{2} \end{cases} \qquad (3.69)$$

Case 2: Over-modulation
In the case of over-modulation, the switching times t'_x, t'_y and t'_0 are simply computed from the original voltage vector and then modified according to the geometrical relationship assuming that $t_{zero} = 0$.

$$t'_x = \frac{t_x}{t_x + t_y + t_0} T_s \qquad (3.70)$$

$$t'_y = \frac{t_x}{t_x + t_y + t_0} T_s \qquad (3.71)$$

Positive Zero Vector is required in:

$$\begin{cases} t_0' = t_{op} = \frac{t_0}{t_x + t_y + t_0} T_s \\ t_{on} = t_{zero} = 0 \end{cases} \tag{3.72}$$

While negative Zero Vector is required in:

$$\begin{cases} t_0' = t_{on} = \frac{t_0}{t_x + t_y + t_0} T_s \\ t_{op} = t_{zero} = 0 \end{cases} \tag{3.73}$$

The control scheme of 3DSVM for a shunt power quality compensator is shown in Fig. 3.44. The determination of the instantaneous reference current for compensation is discussed by the Generalized Instantaneous Reactive Power Theory [14] for a three-phase four-wire system.

A simulation is performed. Figure 3.45 shows the nonlinear load current before the compensation. The current unbalance, harmonic and neutral currents needed to be compensated. Figure 3.46 shows the source currents after compensation. It is obvious that the 3DSVM could control the APF to reduce the harmonic spectrum and compensate the reactive and unbalance currents with eliminating neutral current as well. The switching frequency is 5 kHz. Table 3.20 shows the performance comparison between the Cylindrical Coordinate PWM and the SVM. The results show that the 3DSVM has better performance at the same switching frequency.

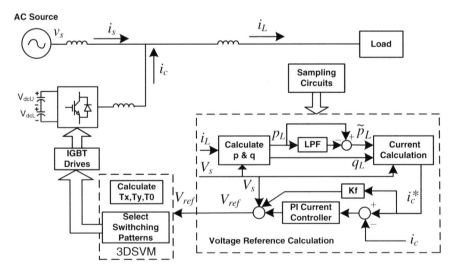

Fig. 3.44 Control scheme of 3DSVM

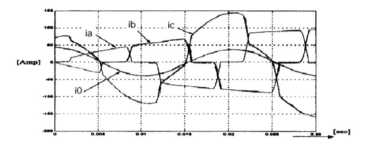

Fig. 3.45 Load current before compensation

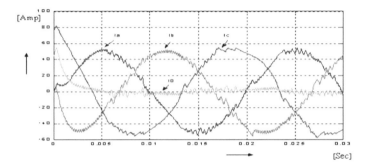

Fig. 3.46 Source current after compensation

Table 3.20 Comparison of cylindrical PWM and SVM performance

Comparison of 3D PWM performance	Without compensation	3D cylindrical coordinate PWM 5 kHz	3D SVM 5 kHz
THD of phase a current (%)	30.47	10.42	6.38
THD of phase b current (%)	32.22	10.01	6.05
THD of phase c current (%)	41.47	6.67	6.89
J_α	15.59	2.64	1.84
J_β	12.13	5.58	3.05
J_0	18.67	2.81	2.39
$J_{\alpha\beta0}$	46.39	9.03	7.28

3.3.5.2 Generalized Three-Level SVM

A 3DSVM is proposed for a three-level three-leg NPC inverter [9]. When the three-level three-leg VSI is applied to a three-phase four-wire system, the neutral wire is connected to the mid-point of the dc bus. In the following, all the voltage vectors are represented in per unit, i.e., normalized by $v_{dcU} = v_{dcL} = v_{dc}/2 = E$, so that the output voltage of each leg has the same value as the switching function S_j. The dc voltage unbalance between the upper and lower capacitor of the three-leg center-split VSI is not considered here, but the dc upper and lower voltage unbalance and its control are discussed in Sect. 3.3.6. The output voltage of each phase can be expressed as:

$$v_x = S_x, x = a, b, c \tag{3.74}$$

A. Decomposition of Reference Voltage Vector

The normalization facilitates the following steps in the 3DSVM. The given reference voltage $\vec{v}_{ref}$ is usually the phase-to-neutral voltage, and the reference voltage vector $\vec{v}_{ref}$ is also normalized by E. A dc offset of half of the dc bus voltage is added to the reference voltages, and the normalized reference voltage vector $\vec{v}_{ref}$ is given in (3.75).

$$\vec{v}_{ref} = \begin{bmatrix} V_{refa}/E \\ V_{refb}/E \\ V_{refc}/E \end{bmatrix} + E \tag{3.75}$$

In this proposed 3D SVM, the desired output voltage vector $\vec{v}_{ref}$ is decomposed to two components as expressed in (3.76) and (3.77):

$$\vec{v}_{ref} = \vec{v}_{offset} + \vec{v}_{twol} \tag{3.76}$$

$$\begin{bmatrix} v_{refa} \\ v_{refb} \\ v_{refc} \end{bmatrix} = \begin{bmatrix} v_{offset(a)} \\ v_{offset(b)} \\ v_{offset(c)} \end{bmatrix} + \begin{bmatrix} v_{towl(a)} \\ v_{towl(b)} \\ v_{towl(c)} \end{bmatrix} \tag{3.77}$$

The offset component $\vec{v}_{offset}$ of the reference voltage is defined as:

$$\vec{v}_{offset} = \begin{bmatrix} v_{offset(a)} \\ v_{offset(b)} \\ v_{offset(c)} \end{bmatrix} = \begin{bmatrix} Int(v_{refa}) \\ Int(v_{refb}) \\ Int(v_{refc}) \end{bmatrix} \tag{3.78}$$

Function Int( ) removes fractional part of the real input data. Consequently, the two-level component $\vec{v}_{twol}$ of the reference voltage is the fractional part.

$$\vec{v}_{twol} = \vec{v}_{ref} - \vec{v}_{offset} = \begin{bmatrix} v_{twol(a)} \\ v_{twol(b)} \\ v_{twol(c)} \end{bmatrix} \tag{3.79}$$

where $0 \leq \vec{v}_{twol(j)} \leq 1$ (j = a, b, c).

If $(v_{offset(a)}, v_{offset(b)}, v_{offset(c)})$ is equal to (S_a, S_b, S_c), the reference voltage vector must be located inside a two-level space vector allocation formed by (S_a, S_b, S_c), (S_{a+1}, S_b, S_c), (S_a, S_{b+1}, S_c), (S_a, S_b, S_{c+1}), (S_{a+1}, S_{b+1}, S_c), (S_{a+1}, S_b, S_{c+1}), $(S_a, S_{b+1},$

Fig. 3.47 Space vector allocation of two-level vectors in the α-β-0 coordinates

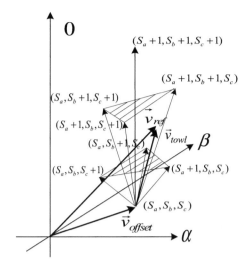

S_{c+1}) and (S_{a+1}, S_{b+1}, S_{c+1}). The two-level space vector allocation in the α-β-0 coordinates is shown in Fig. 3.47 and its projection on the α-β plane is shown in Fig. 3.48. It can be seen from Fig. 3.47 that the vectors (S_{a+1}, S_b, S_c), (S_a, S_{b+1}, S_c) and (S_a, S_b, S_{c+1}) are located on the same horizontal plane, while the vectors (S_{a+1}, S_{b+1}, S_c), (S_{a+1}, S_b, S_{c+1}) and (S_a, S_{b+1}, S_{c+1}) are located on the other horizontal plane.

The decomposition of the reference voltage vector is also illustrated in Figs. 3.47 and 3.48. The 3D parameters of the two-level reference voltage vectors can be expressed as (3.80) according to the α-β-0 transformation.

$$\begin{bmatrix} v_{twol\alpha} \\ v_{twol\beta} \\ v_{twol0} \end{bmatrix} = \sqrt{\frac{2}{3}} \begin{bmatrix} 1 & -1/2 & 1/2 \\ 0 & \sqrt{3}/2 & -\sqrt{3}/2 \\ 1/\sqrt{2} & 1/\sqrt{2} & 1/\sqrt{2} \end{bmatrix} \begin{bmatrix} v_{twol(a)} \\ v_{twol(b)} \\ v_{twol(c)} \end{bmatrix} \tag{3.80}$$

Fig. 3.48 Projection of space vector allocation of two-level vectors on the α-β plane

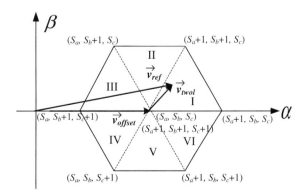

B. Synthesize Reference Voltage Vector

In the 3D SVM, the reference voltage vector is synthesized by four neighboring vectors.

$$\vec{v}_{ref} \cdot T_S = \vec{v}_1 T_1 + \vec{v}_2 T_2 + \vec{v}_3 T_3 + \vec{v}_4 T_4 \tag{3.81}$$

If the dwell time of each vector is normalized by the PWM control period T_s, the following results can be obtained:

$$\vec{v}_{ref} = \vec{v}_1 t_1 + \vec{v}_2 t_2 + \vec{v}_3 t_3 + \vec{v}_4 t_4 \tag{3.82}$$

and $t_1 + t_2 + t_3 + t_4 = 1$

The neighboring vector $\vec{v}_1$ is the offset voltage vector determined by (3.78), which locates at the center point of the eight vectors on the α-β plane. $\vec{v}_2$ and $\vec{v}_3$ are chosen as the neighboring vectors of the reference voltage vector on the α-β plane. It can be seen from Fig. 3.48 that the vectors (S_a, S_b, S_c) and (S_{a+1}, S_{b+1}, S_{c+1}) overlap on the α-β plane. The zero-sequence compensation is achieved by introducing the fourth vector in reference voltage approximation, and $\vec{v}_4$ is always (S_{a+1}, S_{b+1}, S_{c+1}).

There are six sections as the neighboring vectors of the reference voltage vector are different. The switching sequences for each section are listed in Table 3.21. The dwell time of each vector is calculated and listed in Table 3.21 as well. The symmetrically aligned scheme is chosen, because it gives the lowest output voltage distortion and current ripples. The switching sequence of the generalized 3DSVM is $\vec{v}_1 \rightarrow \vec{v}_2 \rightarrow \vec{v}_3 \rightarrow \vec{v}_4$ in the first half period and in reverse turn $\vec{v}_4 \rightarrow \vec{v}_3 \rightarrow \vec{v}_2 \rightarrow \vec{v}_1$ for the next half period. Figure 3.49 illustrates the output modulation signals which correspond to the sequence in Section I in Table 3.21. The boundary conditions of each section in Fig. 3.48 are also given in Table 3.21 and the flow chart of determining the sections is illustrated in Fig. 3.50.

The flow chart of the 3DSVM is shown in Fig. 3.51, which can be applied to three-leg center-split VSIs of different levels without modification. Moreover, the computational cost is always the same. Therefore, the generalized 3DSVM is given above for multi-level three-leg center-split VSI [28].

C. Simulation Results

The system configuration in Fig. 3.27 is used to verify the validity of this generalized 3DSVM. The load currents and the source currents after compensation are shown in Figs. 3.52 and 3.53 respectively. The THD of the source currents reduce from 41.7 to 5.4 %. The sampling frequency is 5 kHz.

Table 3.21 Boundary conditions, neighboring vectors and dwell times in the α-β-0 coordinates

Section	Boundary conditions	Neighboring vectors	Dwell times
I	$v_{twol\beta} > 0$ $v_{twol\alpha} > 0$ $v_{twol\beta} - \sqrt{3}v_{twol\alpha} < 0$	$\vec{v}_1 = (S_a, S_b, S_c)$ $\vec{v}_2 = (S_a + 1, S_b, S_c)$ $\vec{v}_3 = (S_a + 1, S_b + 1, S_c)$ $\vec{v}_4 = (S_a + 1, S_b + 1, S_c + 1)$	$t_1 = 1 - t_2 - t_3 - t_4$ $t_2 = (\sqrt{3}/2)v_{twol\alpha} - 0.5t_3$ $t_3 = \sqrt{2}v_{twol\beta}$ $t_4 = v_{twol0}/\sqrt{3} - t_2/3 - 2t_3/3$
II	$v_{twol\beta} > 0$ $\lvert v_{twol\beta} \rvert - \sqrt{3}\lvert v_{twol\alpha} \rvert > 0$	$\vec{v}_1 = (S_a, S_b, S_c)$ $\vec{v}_2 = (S_a, S_b + 1, S_c)$ $\vec{v}_3 = (S_a + 1, S_b + 1, S_c)$ $\vec{v}_4 = (S_a + 1, S_b + 1, S_c + 1)$	$t_1 = 1 - t_2 - t_3 - t_4$ $t_2 = t_3 - \sqrt{6}v_{twol\alpha}$ $t_3 = \sqrt{3}/2 v_{twol\alpha} + \sqrt{1/2}v_{twol\beta}$ $t_4 = v_{twol0}/\sqrt{3} - t_2/3 - 2t_3/3$
III	$v_{twol\beta} > 0$ $v_{twol\alpha} < 0$ $v_{twol\beta} - \sqrt{3}v_{twol\alpha} < 0$	$\vec{v}_1 = (S_a, S_b, S_c)$ $\vec{v}_2 = (S_a, S_b + 1, S_c)$ $\vec{v}_3 = (S_a, S_b + 1, S_c + 1)$ $\vec{v}_4 = (S_a + 1, S_b + 1, S_c + 1)$	$t_1 = 1 - t_2 - t_3 - t_4$ $t_2 = \sqrt{2}v_{twol\beta}$ $t_3 = (-\sqrt{3}/2)v_{twol\alpha} + 0.5 \cdot t_2$ $t_4 = v_{twol0}/\sqrt{3} - t_2/3 - 2t_3/3$
IV	$v_{twol\beta} < 0$ $v_{twol\alpha} < 0$ $v_{twol\beta} - \sqrt{3}v_{twol\alpha} > 0$	$\vec{v}_1 = (S_a, S_b, S_c)$ $\vec{v}_2 = (S_a, S_b, S_c + 1)$ $\vec{v}_3 = (S_a, S_b + 1, S_c + 1)$ $\vec{v}_4 = (S_a + 1, S_b + 1, S_c + 1)$	$t_1 = 1 - t_2 - t_3 - t_4$ $t_2 = -\sqrt{2}v_{twol\beta}$ $t_3 = (-\sqrt{3}/2)v_{twol\alpha} - 0.5 \cdot t_2$ $t_4 = v_{twol0}/\sqrt{3} - t_2/3 - 2t_3/3$
V	$v_{twol\beta} < 0$ $\lvert v_{twol\beta} \rvert - \sqrt{3}\lvert v_{twol\alpha} \rvert > 0$	$\vec{v}_1 = (S_a, S_b, S_c)$ $\vec{v}_2 = (S_a, S_b, S_c + 1)$ $\vec{v}_3 = (S_a + 1, S_b, S_c + 1)$ $\vec{v}_4 = (S_a + 1, S_b + 1, S_c + 1)$	$t_1 = 1 - t_2 - t_3 - t_4$ $t_2 = t_3 - \sqrt{6}v_{twol\alpha}$ $t_3 = \sqrt{3}/2 v_{twol\alpha} - \sqrt{1/2}v_{twol\beta}$ $t_4 = v_{twol0}/\sqrt{3} - t_2/3 - 2t_3/3$
VI	$v_{twol\beta} < 0$ $v_{twol\alpha} > 0$ $v_{twol\beta} + \sqrt{3}v_{twol\alpha} > 0$	$\vec{v}_1 = (S_a, S_b, S_c)$ $\vec{v}_2 = (S_a + 1, S_b, S_c)$ $\vec{v}_3 = (S_a + 1, S_b, S_c + 1)$ $\vec{v}_4 = (S_a + 1, S_b + 1, S_c + 1)$	$t_1 = 1 - t_2 - t_3 - t_4$ $t_2 = (\sqrt{3}/2)v_{twol\alpha} - 0.5 \cdot t_2$ $t_3 = -\sqrt{2}v_{twol\beta}$ $t_4 = v_{twol0}/\sqrt{3} - t_2/3 - 2t_3/3$

Fig. 3.49 PWM output for one period

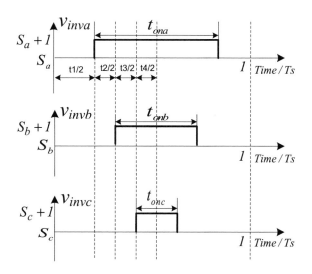

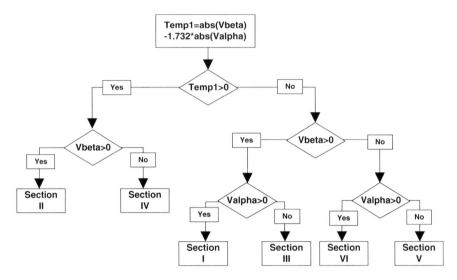

Fig. 3.50 Determination of the sections

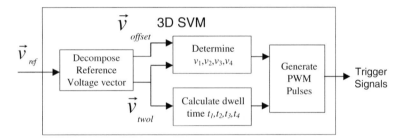

Fig. 3.51 Flow chart of the 3DSVM

3.3.6 DC Linked Voltage Variation Control

3.3.6.1 Analysis of DC Voltage Variation

In the three-leg center-split VSI structure as shown in Fig. 3.54, the ac neutral wire
is connected directly to the mid-point of the two dc linked capacitors. The current of
each phase i_{cx} (x = a, b, c) is forced to flow through either upper capacitor, C_1, or
lower capacitor, C_2, of the dc bus. Thus the relationship between dc voltage variation and three-phase current can be expressed as Eq. (3.83), where the positive
direction of injecting current is as shown in Fig. 3.54 and $C_{dc} = C_1 = C_2$. The result
in this section can be applied into multi-level neutral point clamped inverters.

For simplicity, the dc voltage variation caused by one leg of the center-split VSI
is considered as Eq. (3.84). A difference in direction of current through the upper or

Current A

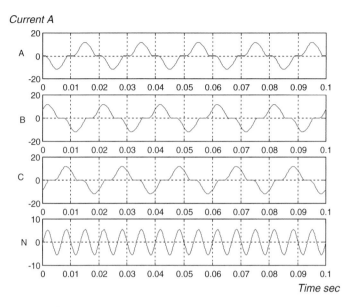

Fig. 3.52 Load currents

Current A

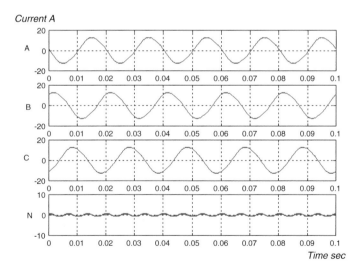

Fig. 3.53 Source current after compensation by 5 kHz three-level 3DSVM

lower capacitor causes dc voltage variation. The relationship between rising or falling current of one leg and dc voltage variation is explained with respect to Fig. 3.55. The dc voltage variations in the dc linked capacitors are summarized in Table 3.22

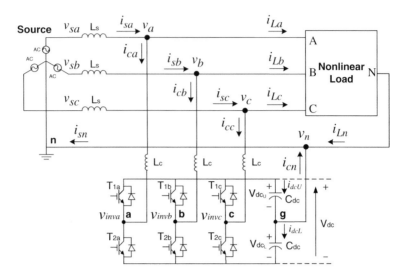

Fig. 3.54 "Center-Split" VSI as a power quality compensator

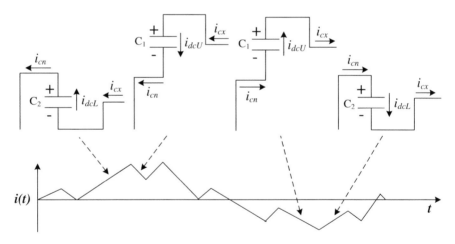

Fig. 3.55 Phase current and capacitor voltages variation

Table 3.22 DC variation conditions for capacitor voltage		
$i_{cx} > 0$ and $\frac{di_{cx}}{dt} < 0$	Increase the voltage in C_1
$i_{cx} < 0$ and $\frac{di_{cx}}{dt} < 0$	Decrease the voltage in C_1
$i_{cx} > 0$ and $\frac{di_{cx}}{dt} > 0$	Decrease the voltage in C_2
$i_{cx} < 0$ and $\frac{di_{cx}}{dt} < 0$	Increase the voltage in C_2

$$C_{dc}\frac{dV_{dcU}}{dt} + C_{dc}\frac{dV_{dcL}}{dt} = S_a \cdot i_{ca} + S_b \cdot i_{cb} + S_c \cdot i_{cc} \qquad (3.83)$$

$$\begin{cases} \frac{dV_{dcU}}{dt} = \frac{1}{C_{dc}}i_{cx} & when \quad S_x = 1 \\ \frac{dV_{dcL}}{dt} = \frac{-1}{C}i_{cx} & when \quad S_x = -1 \end{cases} \quad x = a, b, c \qquad (3.84)$$

In the following analysis, S_x is defined as the switching function where the effect of dc voltage variation is not considered. And S_x^N is the new switching function where the dc voltage variation is considered. When $V_{dcU} \neq V_{dcL}$ is employed, E can be redefined as $E = (V_{dcU} + V_{dcL})/2$. Therefore, the corresponding new switching function is described as:

$$S_x^N = \begin{cases} V_{dcU}/E & upper leg\ is\ on \\ -V_{dcL}/E & lower\ is\ on \end{cases} \quad x = a, b, c \qquad (3.85)$$

And

$$S_x^N = \begin{cases} \frac{V_{dcU}}{E} = \frac{2V_{dcU}}{V_{dcU} + V_{dcL}} = \frac{V_{dcU} + V_{dcL} + V_{dcU} - V_{dcL}}{V_{dcU} + V_{dcL}} = 1 + \frac{V_{dcU} - V_{dcL}}{V_{dcU} + V_{dcL}} \\ -\frac{V_{dcL}}{E} = -\frac{2V_{dcL}}{V_{dcU} + V_{dcL}} = \frac{-(V_{dcU} + V_{dcL}) + V_{dcU} - V_{dcL}}{V_{dcU} + V_{dcL}} = -1 + \frac{V_{dcU} - V_{dcL}}{V_{dcU} + V_{dcL}} \end{cases} \qquad (3.86)$$

If $\Delta S = (V_{dcU} - V_{dcL})/(2E)$ is substituted to Eq. (3.86), it can be simplified as:

$$S_x^N = \begin{cases} 1 + \Delta S \\ -1 + \Delta S \end{cases} \quad x = a, b, c \qquad (3.87)$$

Comparing Eq. (3.87) with conventional switching function, the relation between the new switching function and the original one can be expressed as:

$$S_x^N = S_x + \Delta S \qquad (3.88)$$

Hence, the output voltage of the inverter can be recalculated by these new S_a^N, S_b^N and S_c^N. The instantaneous voltage vector $\vec{V}^N$ in 3D, α-β-0 coordinate, can be expressed as:

$$\vec{v}^N = \frac{v_{dc}}{2}\left[\sqrt{\frac{2}{3}}S_\alpha^N \cdot \vec{n}_\alpha + \frac{1}{\sqrt{2}}S_\beta^N \cdot \vec{n}_\beta + \frac{1}{\sqrt{3}}S_0^N \cdot \vec{n}_0\right] \qquad (3.89)$$

where

$$S_\alpha^N = S_a^N - \frac{1}{2}S_b^N - \frac{1}{2}S_c^N \qquad (3.90)$$

$$S_\beta^N = S_b^N - S_c^N \qquad (3.91)$$

$$S_0^N = S_a^N + S_b^N + S_c^N \tag{3.92}$$

Substituting the new switching function, (3.88), into (3.90), (3.91) and (3.92),

$$S_\alpha^N = S_\alpha \tag{3.93}$$

$$S_\beta^N = S_\beta \tag{3.94}$$

$$S_0^N = S_0 + 3\Delta S \tag{3.95}$$

As a result, only zero-frame reference of the original 3D vector's allocation is changed under the dc voltage variation situation. And only the compensation performance of the zero-frame is affected accordingly. However, when the output voltage is reversely transformed to *a-b-c* frame, current compensation in each phase will be affected. Therefore the issues of dc voltage variation must be controlled in practical applications. Hereinafter, the dc voltage variation control strategy [27] is given.

3.3.6.2 DC Voltage Variation Control Strategy

In the space vector's allocation which is shown in Fig. 3.29, there are two switching patterns for the zero voltage vectors: one positive $\vec{V}_{op}$ and one negative $\vec{V}_{on}$. The α and the β axes components of $\vec{V}_{op}$ and $\vec{V}_{on}$ are equal to zero. According to the results obtained in last part, the dc voltage variation affects only the zero-frame reference. Hence, dc voltage variation is controlled by varying the switching times of $\vec{V}_{op}$ and $\vec{V}_{on}$.

Figure 3.56 shows the current direction when the vector $\vec{V}_{op}$ or $\vec{V}_{on}$ is chosen. It is obvious that the inverter's current only passes through one of the dc capacitor when the voltage vector is $\vec{V}_{op}$ or $\vec{V}_{on}$. When the output vector of the inverter is $\vec{V}_{op}$, which corresponds to the switching pattern (1, 1, 1), the upper capacitor C_1 is

(a)   **(b)**

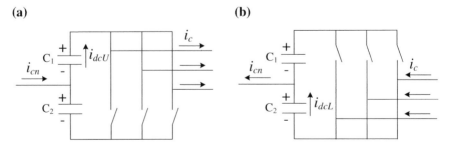

Fig. 3.56 Phase current under zero switching pattern: **a** Switching pattern (1, 1, 1) and **b** switching pattern (−1, −1, −1)

discharged. Reversely, if the output vector is $\vec{V}_{on}$, corresponding to the switching pattern $(-1, -1, -1)$, the lower capacitor C_2 is discharged. Therefore $\vec{V}_{op}$ and $\vec{V}_{on}$ can change the dc voltage imbalance in opposite direction.

In the above-mentioned 3DSVM control technique, $\vec{V}_{op}$ and $\vec{V}_{on}$ share the same dwell time $t_{zero}/2$ to approximate for $V_{zero} \cdot t_{zero} = 0$. Hence, if the dwell time of $\vec{V}_{op}$ and $\vec{V}_{on}$ are varied, the output of $V_{zero} \cdot t_{zero}$ can not be equal to zero. However, since the α and the β axis components of $\vec{V}_{op}$ and $\vec{V}_{on}$ are all zeros, the $V_{zero} \cdot t_{zero}$ only has a zero sequence output. In the proposed dc variation control strategy, a new variable, $\varepsilon (0 < \varepsilon < 1)$, is introduced to vary the dwell time of $\vec{V}_{op}$ and $\vec{V}_{on}$, i.e., the dwell time of $\vec{V}_{op}$ is assigned to be $(1 + \varepsilon)t_{zero}/2$, and thereby the dwell time of $\vec{V}_{on}$ automatically takes $(1 - \varepsilon)t_{zero}/2$. A different ε value corresponds to a different $V_{zero} \cdot t_{zero}$ output, which is employed to balance the neutral point voltage. This method is often used to balance the dc voltage of the three-level neutral-point-clamped inverter.

Furthermore the larger the value of ε, the higher is the dc control ability. So it is reasonable to consider choosing ε as '1' or '−1' to maximize the ability of dc voltage variation control in the time range of t_{zero}. The relation of dc voltage variation and ε values are listed in Table 3.23, where k is the proportional coefficient with $0 < k < 1$ and V_{dc} is the reference dc voltage. The 'deltav' can be calculated by the expression described in Eq. (3.96). However, there is a symbol '*' which means "cannot be defined" in Table 3.23 when the "deltav" exceeds the maximum dc variation limitation V_{max}. If the detected dc voltage imbalance is too large, it will affect the stability of the system. In that case, the control strategy in Table 3.24 can be considered, in which the whole period of sampling time T_s is employed to control the dc voltage variation.

$$deltav = abs(v_{dcU}) - abs(v_{dcL}) \qquad (3.96)$$

Table 3.23 Relation of dc voltage and ε

Cases	Condition	ε
Serious dc voltage imbalance	$abc(deltav) > V_{max}$	*
Largest lower voltage imbalance	$-V_{max} < deltav < -k \cdot V_{dc}$	−1
Larger lower voltage imbalance	$-k \cdot V_{dc} < deltav < 0$	$deltav/(kV_{dc})$
Larger upper voltage imbalance	$0 < deltav < k \cdot V_{dc}$	$deltav/(kV_{dc})$
Largest upper voltage imbalance	$k \cdot V_{dc} < deltav < V_{max}$	+1

Table 3.24 Time distribution under severe dc imbalance case

Condition	Dwell time of $\vec{V}_{0p}$	Dwell time of $\vec{V}_{0n}$
$deltav > V_{max}$	T_S	0
$deltav < -V_{max}$	0	T_S

3.3.6.3 Simulation Results

In this part, simulation is performed by Matlab/Simulink. The system configuration is shown in Fig. 3.54. The 3DSVM control strategy is implemented with 2.5 kHz switching frequency. The three-phase and the neutral currents at the load side are shown in Fig. 3.57. When the proposed 3DSVM without controlling the dc voltage variation is applied, the harmonic and neutral currents are compensated simultaneously as shown in Fig. 3.58. Particular loads are chosen in this simulation, for which the system neutral current is always positive without compensation. Thus there is higher possibility of neutral current to pass through one side of the inverter's dc bus, e.g. the upper capacitor. As a result, the voltage of one capacitor continues to increase while the voltage of the other one continues to decrease. Large dc voltage unbalance occurs under this situation, which is shown in Fig. 3.59.

Figures 3.60 and 3.61 show the compensation results when the 3DSVM with dc voltage variation control is implemented. The control strategy mentioned in Tables 3.23 and 3.24 are employed to control the dc voltage variation. The capacitor voltage variation after being controlled is shown in Fig. 3.60. In the simulation, following parameters are used: system ac voltage RMS value = 70 V, capacitor dc voltage V_{dc} = 100 V, k = 10 %, V_{max} = 15 V. In Fig. 3.60, there are obviously three periods of capacitor voltage variation which correspond to the three steps of the control strategies mentioned in Tables 3.23 and 3.24. Figure 3.61 is the source current compensation result where the dc voltage imbalance is controlled simultaneously. However, compared to Fig. 3.58, the currents after compensation

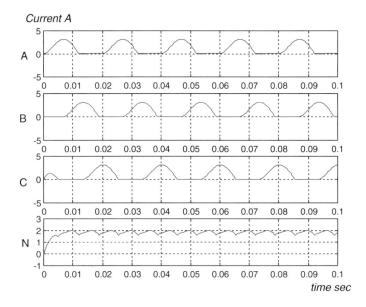

Fig. 3.57 Load current

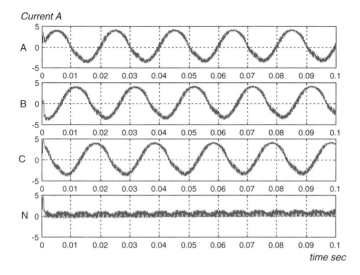

Fig. 3.58 Source current after compensation by using 2.5 kHz 3DSVM

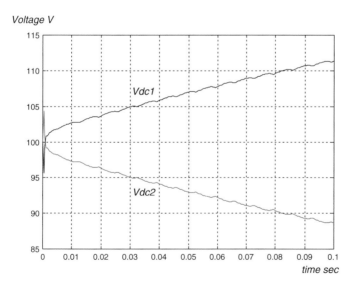

Fig. 3.59 dc voltage without being controlled

are influenced. It is mainly due to the fact that when the dc voltage variation is larger than the maximum limitation V_{max}, the switching time is dedicated to control the dc imbalance instead of the power quality compensation according to the control strategy in Table 3.24.

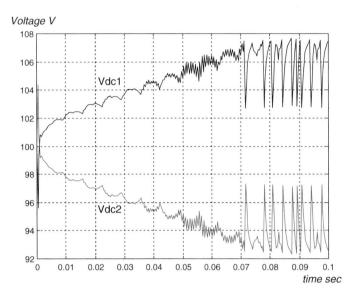

Fig. 3.60 DC Voltage after being controlled

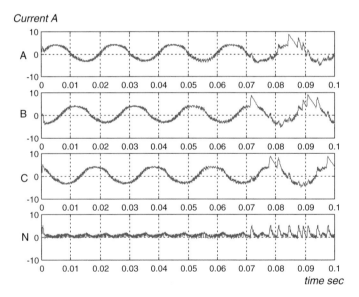

Fig. 3.61 Source current after d.c. voltage variation control is added

In simulation, the extreme case is chosen in order to explain the dc voltage variation control strategy more clearly. All the control strategy mentioned in Tables 3.23 and 3.24 are tested. Since the triple harmonic components dominates the waveform of the neutral current in distribution site from statistical results, the strategy mentioned in Table 3.23 is enough for the dc voltage variation control in most practical applications. In addition, the situation discussed in Table 3.24 can also be avoided by increasing the size of the dc capacitor so that better power quality compensating performance is obtained.

3.4 Three-Phase Four-Wire Multi-Level VSIs

In high power applications, multi-level topologies can be considered as one of the solutions. Multi-level topologies have the following advantages: reduced harmonic content of output waveform and reduced voltage stress across switching components. In this section, it is dedicated for the discussion of choosing a suitable circuit topology among the three-phase converters for high power applications in the three-phase four-wire systems. For simplicity, only the three-level systems are chosen as an example for discussion, but the analyzed results can be applied in other higher-level applications.

Figure 3.62 shows a three-level three-leg VSI topology in a three-phase three-wire system, it is obvious that it cannot supply zero voltage and neutral current so that this topology is not considered in a three-phase four-wire application. Basically, a three-leg center-split VSI topology in a three-level case can be considered with a neutral wire connection with "O" and "n" nodes as shown in Fig. 3.63. When a neutral wire of a three-phase four-wire system is connected with the mid-point of dc capacitors in a three-level NPC inverter, it turns into a

Fig. 3.62 A three-level three-leg VSI Topology

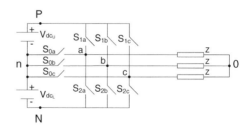

Fig. 3.63 A three-level three-leg center-split VSI topology

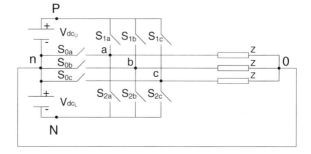

three-phase four-wire system. Without neutral wire connection, a three-level NPC inverter is operated in a three-phase three-wire case. In short, it means that the same three-level topology can be employed for three-phase three-wire and four-wire systems in multi-level converter applications.

Furthermore, a three-level four-leg VSI topology is under consideration, as shown in Fig. 3.64. In Fig. 3.64, when S_{Of} is turned off, it does not affect the operation of the neutral current. When $S_{Of} = on$, it has two parallel neutral current paths, which can be considered as one path connecting between "f" and "n" nodes. As a result, the circuit turns into a three-level three-leg center-split VSI with an additional leg as shown in Fig. 3.65. Both circuit topologies as shown in Figs. 3.64 and 3.65 can provide neutral current control ability. Although the multi-level four-leg topology can give more degree of freedom in selection of available switching states, it is not worth to have it in a practical system due to the higher initial cost with an additional leg in the view point of the neutral current control and compensation. In addition, the control of the upper-leg and the lower-leg dc capacitor voltages is necessary in both three-level three-leg center-split and three-level four-leg topologies. Based on the results in Chap. 2, the three-leg center-split and the four-leg topologies provide the same $\alpha\beta$ voltage capacity. Due to the considerations of the initial cost and the same $\alpha\beta$ voltage capacity, the three-leg center-split topology is seen as the most convenient structure for the development of the three-phase four-wire multi-level VSIs in high power applications.

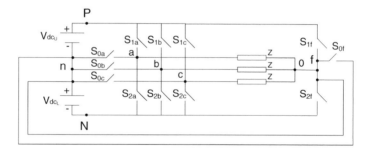

Fig. 3.64 A three-level four-leg VSI topology with S_{Of} turned off

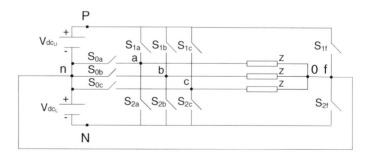

Fig. 3.65 A three-level four-leg VSI topology with S_{Of} turned on

3.5 Generalized PWM for Multi-Level Three-Leg Center-Split and Four-Leg VSIs

3.5.1 Generalized 3D Direct PWM

The generalized 3D direct PWM [20, 29] is based on the volt-second approximation as the conventional space vector modulation. It can be used to control the three-leg center-split VSIs and the four-leg VSIs from the two-level to the multi-level topologies [30, 31], as illustrated in Fig. 3.66. The generalized 3D direct PWM implements the voltage synthesis in per-leg mode. The pulse width of each phase is directly determined to track given voltage references.

In order to deduce the 3D direct PWM, an equivalent model for an N-level three-leg center-split VSI is introduced in Fig. 3.67.

The output voltage of each inverter leg is expressed as:

$$V_{nvx} = E \cdot S_x \quad x = a, b, c \tag{3.97}$$

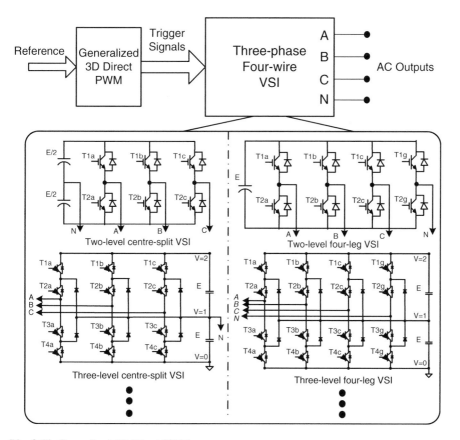

Fig. 3.66 Generalized 3D Direct PWM

Fig. 3.67 Equivalent model
of a three-leg N-level voltage
source inverter

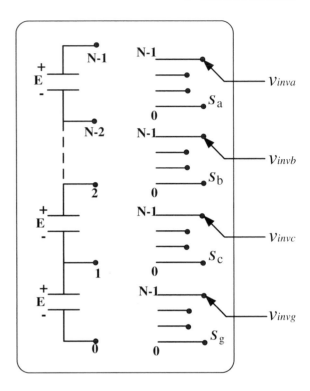

where E is the voltage of one level, and Sx is the switching function. For an N-level
inverter, the value of S_j varies among 0 to N − 1, and the dc bus voltage equals
(N − 1)E. When the N-level three-leg inverters are applied to a three-phase
four-wire system, the neutral wire is connected to the mid-point of the dc bus.
Hence, the voltage vg in reference to the virtual ground in Fig. 3.67 is expressed as:

$$V_g = E(N-1)/2 \tag{3.98}$$

All the voltages are expressed per unit, i.e., normalized by E, so that the output
voltage of each leg has the same value as the switching function S_j.

$$v_{invj} = S_j \quad j = a, b, c \tag{3.99}$$

and

$$v_{invg} = (N-1)/2 \tag{3.100}$$

The flow chart of the generalized 3D direct PWM is shown in Fig. 3.68. The
shifting voltage must be used to determine the reference of each leg when four-leg
inverters are controlled. A mode signal for indicating the inverter topology is
needed when the shifting voltage is used, as illustrated in Fig. 3.68.

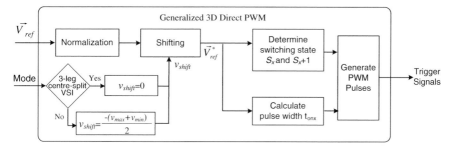

Fig. 3.68 Flow chart of the generalized 3D direct PWM

The detailed procedures for implementing the 3D direct PWM are provided hereinafter.

(1) The input reference voltages are first normalized by voltage of one level.

$$\vec{v}_{refx} = \begin{bmatrix} v_{refa}/E \\ v_{refb}/E \\ v_{refc}/E \end{bmatrix} \tag{3.101}$$

(2) The reference voltage of each leg is then calculated.

$$v^*_{refx} = v_{refx} + v_{shift} \quad x = a, b, c, g \tag{3.102}$$

The shifting voltage is defined as:

$$v_{shift} = \begin{cases} 0 & 3 - leg \quad centre-split \quad VSI \\ -(v_{max} + v_{min})/2 & Otherwise \end{cases} \tag{3.103}$$

where v_{max} is defined as the maximum voltage among v_{refa}, v_{refb}, v_{refc} and v_{refg}, v_{min} is the minimum voltage among g v_{refa}, v_{refb}, v_{refc} and v_{refg}, and v_{refg} always equals to zero.

(3) The switching states of each leg are determined by

$$S_x = Int\left(v^*_{refx} + \frac{N-1}{2}\right) \quad x = a, b, c, g \tag{3.104}$$

The function **Int** removes all the fractional part from the input value. According to (3.104), the reference voltage is a value between S_x and S_{x+1}. As a result, the output voltage of this leg is synthesized by two switching states, S_x and S_{x+1}, in one switching period.

(4) The dwell time of each switching state is then calculated as given in (3.105). The fractional part is just the normalized pulse width, i.e., the dwell time of S_{x+1}.

Fig. 3.69 Switching interval
of one inverter leg

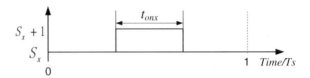

$$t_{onx} = v_{refx}^* + \frac{N-1}{2} - S_x \quad x = a, b, c, g \tag{3.105}$$

(5) According to the switching states and dwell times, the switching patterns are
    determined and the final trigger signals of each switch of one inverter leg are
    generated, as shown in Fig. 3.69.
    Simulations are implemented by Matlab/Simulink. The simulation system
    configuration is shown in Fig. 3.70. The output voltages of the VSI passed
    through an output filter and are applied to three-phase balanced loads. The
    filter is designed to smooth the pulses and recover the average analog wave-
    forms. The 3D direct PWM is applied in order to control the three-phase
    four-wire VSIs to track reference voltages.

The reference voltages are selected to illustrate that both pure sinusoidal and
harmonic-injection phase-to-neutral reference voltages can be generated by using
the 3D direct PWM. A two-level center-split VSI, a three-level center-split VSI, a
five-level center-split VSI, a two-level four-leg VSI and a three-level four-leg VSI
are controlled respectively. The reference and output voltages of each leg of the
corresponding VSIs are shown in Figs. 3.71, 3.72, 3.73, 3.74, 3.75.

The voltages after passing the filter and currents through the loads in a
three-level center-split VSI are shown in Fig. 3.76. Similar results were obtained in
the four other cases. The THD and RMS values of the output voltages after passing
the filter are provided in Table 3.25. The simulation results indicate that the 3D
direct PWM can control the three-leg center-split VSI and four-leg VSI from
two-level to multi-level topologies.

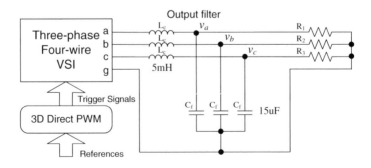

Fig. 3.70 Simulation system configuration

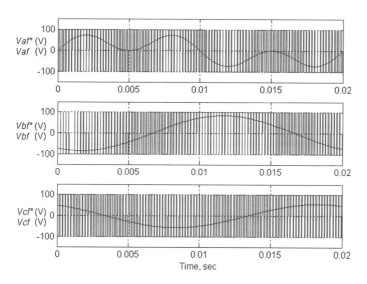

Fig. 3.71 Output voltage of a two-level three-leg center-split VSI

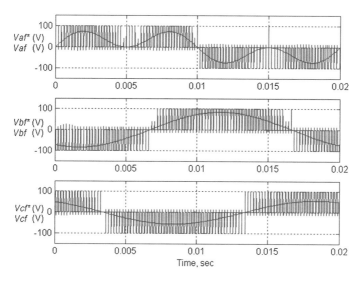

Fig. 3.72 Output voltage of a three-level three-leg center-split VSI

3.5.2 Generalized FPGA-Based 3D PW Modulator

Based on the 3D direct PWM, a generalized Pulse Width (PW) modulator can be developed by using a XILINX XC3S400 FPGA. The FPGA provides benefits, such as high operating frequency, parallel processing capabilities and user definable I/O ports.

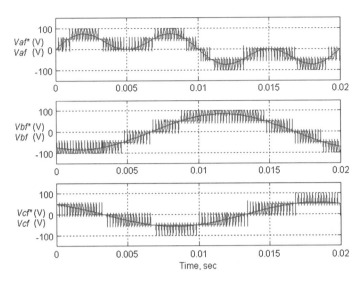

Fig. 3.73 Output voltage of a five-level three-leg center-split VSI

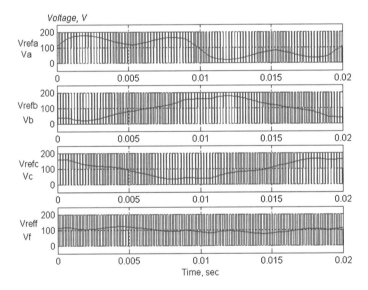

Fig. 3.74 Output voltage of a two-level four-leg VSI

The frequency of the external clock was 40 MHz, and 19 % slices of the FPGA were used. The block diagram of the generalized PW modulator is shown in Fig. 3.77, which consists of three modules.

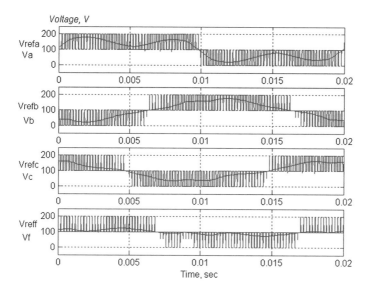

Fig. 3.75 Output voltage of a three-level four-leg VSI

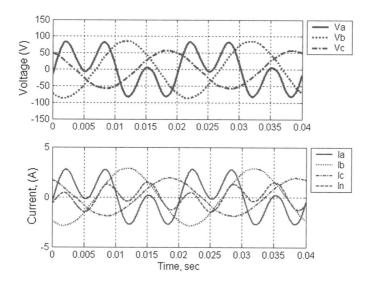

Fig. 3.76 Voltages across the loads and the currents passing through the loads

3.5.2.1 Data Buffer

The I/O port of the PW modulator consists of a 12-bit data bus and a 3-bit address bus. The phase-to-neutral reference voltages are modified according to. (3.106), before being transferred to the FPGA in every sampling period.

Table 3.25 Parameters of simulation results

	THD (%)			RMS (V)		
	A	B	C	A	B	C
Reference voltages	100	1.0	1.0	48.08	60	40
Two-level center-split VSI	105	1.71	2.75	49.58	60.34	40.24
Three-level center-split VSI	105	1.13	1.70	49.53	60.3	40.18
Five-level center-split VSI	105	0.67	1.11	48.99	59.86	39.51
Two-level four-leg VSI	104.9	2.37	2.44	49.58	60.33	40.25
Three-level four-leg VSI	103.2	2.0	2.8	49.01	59.74	40.01

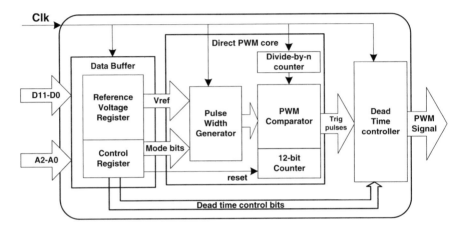

Fig. 3.77 Block diagram of the generalized pulse width modulator

$$\vec{v}_{ref} = INT \left[500 \times \begin{bmatrix} V_{refa}/E \\ V_{refb}/E \\ V_{refc}/E \end{bmatrix} \right] + 250(N-1) \qquad (3.106)$$

where N is the level of the VSI and E is the dc voltage of one level. The floating-point arithmetic becomes fixed-point arithmetic via (3.106).

The period of the digital counter is set to 1000 and its control clock period is 0.2 us, when the switching frequency is 5 kHz. The period of the control clock is much shorter than the turn-off time of most power switches. For example, the maximum turn-off time of the IGBT used in prototypes is 3 us. Hence, representing the pulse widths by an unsigned integer is acceptable when the PW modulator is designed. In addition, 10-bit A/D converters are used in prototypes. When a five-level VSI is controlled, the reference signal sending to the FPGA is in the range of 0–2000. Hence, 12-bit integer realization was selected for the design of the PW modulator.

The data buffer also contains a 12-bit control register which needs to be initialized before the PW modulator is triggered to operate. The detailed description of the control bits are provided hereinafter, in which Bit 0 represents the least significant bit and Bit 11 is the most significant bit.

- Bit 0–1 **Level**: The level of the inverters to be controlled

Bit 1	Bit 0	Inverter Level N
0	0	2
0	1	3
1	0	4
1	1	5

- Bit 2–4 **ICP**: Input clock prescaler for the PWM core

000	F/2	100	F/32
001	F/4	101	F/64
010	F/8	110	F/128
011	F/16	111	F/256

  F = FPGA input clock frequency

- Bit 5 **Mode**: The topology of the inverter

0	Three-leg center-split VSI
1	Four-leg VSI

- Bit 6 **Reset**: Reset the PWM core

0	Reset entire PWM core(counter is set back to 0)
1	No effect

- Bit 7 **Handshaking**

0	The reference data in registers are ready.
1	The reference data in registers are being modified.

- Bit 8-10 **ICPD**: Input Clock Prescaler for the Dead-time (ICPD) controller. Scaling table is the same as that for **ICP**.
- Bit 11 Reserved

3.5.2.2   Direct PWM Core

The direct PWM core contains a divide-by-n counter which scales the external input clock to provide clock signals for comparators and 12-bit counters. The 12-bit up-down counter is triggered by the rising-edge of the 'Reset' bit in the control register. The counter first operates in up-counting mode, counting from 0 to 500. The counter is then in down-counting mode until its output reaches 0. The final PWM output frequency can be determined by:

$$f = F/r_c/1000, \tag{3.107}$$

where F is the frequency of the external clock and r_c is the scaling ratio of the divide-by-n counter which is determined by the value of Input Clock Prescale (ICP) in the control register.

The pulse width generator calculates the pulse width of each switch according to the states of control bits 'Level' and 'Mode'. The detailed steps are as follows:

(1) The reference signals are read from the registers and the reference voltages of each leg are calculated according to (3.108), where V_{refg} always equals 250 (N − 1).

$$V_{invx} = V_{refx} + V_{shift} - 250 \times (N - 1) \quad x = a, b, c, g \tag{3.108}$$

where

$$V_{shift} = \begin{cases} 500 \times (N - 1) - \frac{V_{max} + V_{min}}{2} & Mode = 1 \\ 0 & Mode = 0 \end{cases} \tag{3.109}$$

(2) Calculate the pulse width of each switch of the inverter.

In the proposed 3D direct PWM, the output pulse of each leg needs to be transferred to the trigger signals of each switch of a VSI. Conventional SVM solves this problem by means of a switching table. The switching table for a three-level VSI is shown in Table 3.26 in which the states of all switches of a leg corresponding to an output voltage are listed.

As the level of the VSI increases, the switching table becomes more complicated and takes up more memory space. A method is given for the design of the generalized PW modulator in which the pulse width of each switch is directly determined by the reference voltages. For an N-level VSI, there are 2(N − 1) switches in each leg. The 2(N − 1) switches are numbered according to the following rule:

Table 3.26 Switching table of a three-level VSI

Voltage	S_x	S_{x1}	S_{x2}	S_{x3}	S_{x4}
0	0	Turn-off	Turn-on	Turn-off	Turn-on
E	1	Turn-on	Turn-off	Turn-off	Turn-on
2E	2	Turn-on	Turn-off	Turn-on	Turn-off

All the switches above the point where the output voltage is connected are numbered by odd values from S_{x1}, S_{x3}... to $S_{x(2N-3)}$, while all the switches below that point are numbered by even values from $S_{x(2N-2)}$, $S_{x(2N-4)}$... to S_{x2} (x = a, b, c, g).

Therefore, the 2(N − 1) switches of one leg can be divided into (N − 1) pairs. If the switch pairs are numbered by odd value 'n' (0 ≤ n ≤ 2 N − 3), the mth pair consists of two switches, which are S_{xm} and $S_{x(m+1)}$. The states of the two switches of the same pair are always opposed. Hence, if the trigger signal for the switch S_{xm} is determined, the trigger signal for its counterpart $S_{x(m+1)}$ can be generated by reversing S_{xm}.

After the switching table was analyzed, it was found that the states of all the odd-numbered switches could be determined according to the value of the switching function S_x. It is assumed that S_x equals K. All the odd-numbered switches whose number is larger than (2K − 1) are turned off, and all the odd-numbered switches whose number is less than or equal to (2K − 1) are turned on. In the 3D direct PWM, the output switching state of each leg varies between S_j and S_{j+1} during one PWM period. Hence, the switches from S_{x1} to $S_{x(2K-1)}$ are turned on during the whole period. Switches of which the numbers are larger than (2K + 1) are always turned off. Switch $S_{x(2K+1)}$ is first turned off, and then turned on when the output state is changed to S_{x+1}. The pulse width of S_{x+1} is simply the pulse width of the trigger signal of switch $S_{x(2K+1)}$. Therefore, the pulse width of each switch can be determined by (3.110).

$$T_x = 250(N - m) + (V_{invx} - 250(N - 1)) = 250(1 - m) + V_{invx} \quad x = a, b, c, g$$

(3.110)

If the symmetrical aligned scheme is adopted in generating the trigger signals, the value, which is compared against the output of the counter, is expressed as:

$$T_{xI} = (500 - T_x)/2 = 250(m + 1) - V_{invx} \quad x = a, b, c, g \qquad (3.111)$$

The T_{xnI} is compared to the output of the 12-bit counter at every falling edge of the clock signal. When the output of the counter is larger than the T_{xnI} of the mth switch, the trigger signal of this switch is forced into high mode. Otherwise, the trigger signal is low. The switching state of the switch with the number m + 1 is obtained by reversing that of the mth switch. Generation of the trigger signals of the four switches of one leg of a three-level inverter is illustrated in Fig. 3.78.

In this case, T_{a3I} and T_{aII} are calculated by (3.111) and compared to the output of the counter to determine trigger signals of the switches S_{a1} and S_{a3}. The trigger signals of switches S_{a2} and S_{a4} are obtained by reversing that of S_{a1} and S_{a3} respectively.

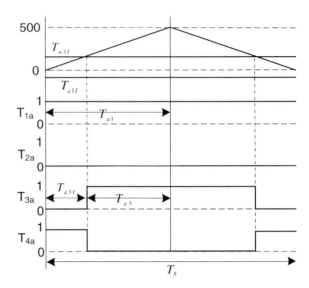

Fig. 3.78 Generation the trigger signals for a three-level VSI

3.5.2.3 Dead-Time Controller

The dead-time controller is included in the generalized PW modulator so that dead-times fit into the transition edges of the trigger signals. The input clock of the dead-time controller is first scaled according to the control bit 'ICPD'. The corresponding clock that is obtained is used as the clock signal for a counter which periodically counts from '0' to '40'. Hence, the dead-time is calculated by (3.112).

$$T_d = 40/(F/r_d), \tag{3.112}$$

where F is the input frequency of the external clock and r_d is the scaling ratio.

3.5.3 Experimental Verification of the 3D PW Modulator

Prototypes for a two-level center-split VSI, a two-level four-leg VSI and a three-level center-split VSI were built. The FPGA-based generalized PW modulator was applied to control these VSIs to track the given references. The block diagram for the control system is shown in Fig. 3.79. A DSP (TMS320F2407) was used to generate the reference voltages and send the references to the data buffers in FPGA.

The RMS value and THD of the references, which are generated by the 16-bit fixed point DSP, are listed in Table 3.27. The dc bus voltage is 200 V and PWM frequency is 5 kHz. The experimental system configuration is the same as that in Fig. 3.70. The experimental results of controlling a two-level center-split VSI, a three-level center-split VSI and a two-level four-leg VSI are shown in Figs. 3.80,

Fig. 3.79 Block diagram for the control system

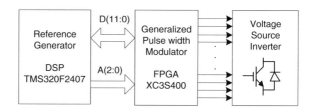

Table 3.27 Parameters of experimental results

	THD (%)			RMS (V)		
	A	B	C	A	B	C
References	100	1.7	1.7	50	44.1	35.35
Two-level center-split VSI	104.8	3.1	2.1	48.6	41.06	29.36
Three-level center-split VSI	104.8	2.5	3.3	50.2	41.76	27.50
Two-level four-leg VSI	102.8	4.8	6.8	50.2	43.2	28.75

3.81 and 3.82, in which the output voltages of the VSIs, the voltage across the loads and the current passing through the loads are provided. Neutral currents are also recorded.

The RMS and THD of the voltages across the loads are provided in Table 3.27. The experimental results show that the FPGA-based generalized PW modulator can control the three-phase four-wire VSIs to trace the given references. In the experiment, the inserted dead-time affects the performance of the four-leg VSI. The performance of the two-level four-leg VSI is deteriorated when the reference voltage of the fourth-leg is low compared to the dc bus voltage. Better performance can be obtained when the modulation index is increased.

3.6 Design and Implementation of Active Power Filters

3.6.1 Minimum DC-Link Voltage Study for APF Under Reactive Power and Current Harmonics Compensation

In the following section, the minimum inverter capacity design for APF under reactive power and current harmonics compensation are discussed. First, the single-phase fundamental and harmonic equivalent circuit models of a three-phase four-wire center-split APF are introduced. According to the current quality of the loading and APF single-phase equivalent circuit models, the minimum dc-link voltage expression for the APF is proposed. Finally, representative simulation results of the three-phase four-wire center-split APF are given to verify its minimum dc-link voltage design expression. Given that most of the loadings in the

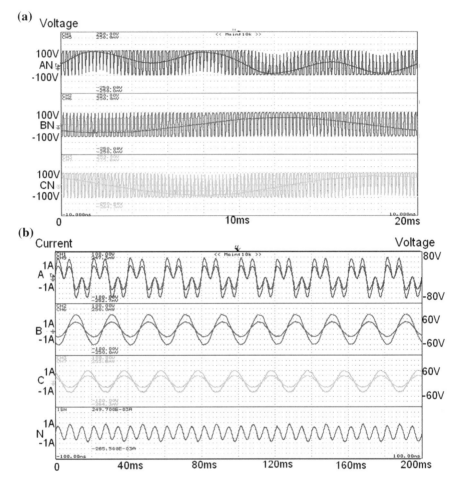

Fig. 3.80 Experimental results of a two-level center-split VSI: **a** Output voltages of the VSI, **b** load voltages and currents

distribution power systems are inductive, the following analysis and discussion only focus on inductive nonlinear loads.

3.6.1.1 Single-Phase Fundamental and Harmonic Equivalent Circuit Models of a Three-Phase Four-Wire APF

Figure 3.83 shows the circuit configuration of the three-phase four-wire center-split APF, where the subscript 'x' denotes phase a, b, c, g. v_{sx} and v_x are the system and load voltages, i_{sx}, i_{Lx}, and i_{cx} are the system, load and inverter currents for each phase. L_c is the coupling inductor of the APF. C_{dc}, V_{dcU}, and V_{dcL} are the dc

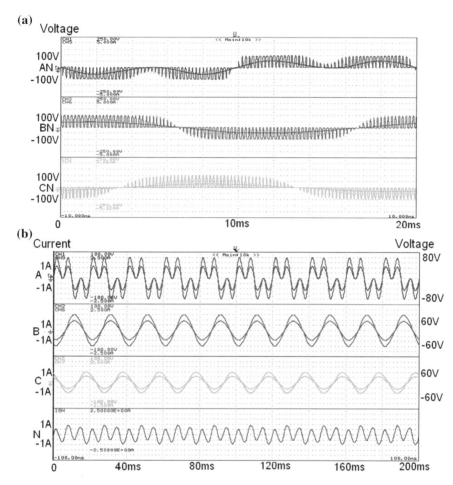

Fig. 3.81 Experimental results of a three-level center-split VSI: **a** Output voltages of the VSI, **b** load voltages and currents

capacitor, upper and lower dc capacitor voltages respectively. In order to simplify the minimum dc-link voltage deduction for the APF, v_{sx} is assumed to be sinusoidal without harmonic component. Moreover, L_s is normally neglected due to its low value relatively, thus $v_{sx} \approx v_x$. Figure 3.84 shows the APF single-phase fundamental and harmonics equivalent circuit models, where the subscript 'g' and 'n' denote the fundamental and harmonics frequency components, and $n = 2, 3\ldots\infty$. Through these two circuit models, the APF minimum dc-link voltage expression with respect to different loading current quality parameters can be deduced. In the following analysis, all parameters are in RMS value.

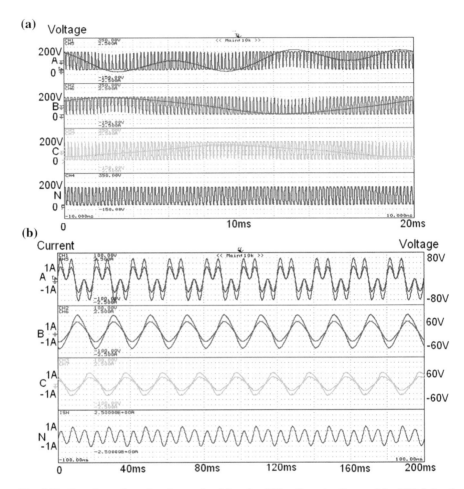

Fig. 3.82 Experimental results of a two-level four-leg VSI. **a** Output voltages of the VSI. **b** Load voltages and currents

3.6.1.2 Minimum DC-Link Voltage Deduction for the Three-Phase Four-Wire APF

Referring to Fig. 3.84b, the required inverter fundamental output voltage (V_{invxf}) and inverter harmonic output voltage (V_{invxn}) at each harmonic order can be found. As V_{invxf} and V_{invxn} are in RMS values, the minimum dc-link voltage values (V_{dcxf}, V_{dcxn}) for compensating the phase fundamental reactive current component and each nth order harmonic current component are calculated as the peak values of the required inverter fundamental and each nth order harmonic output voltages, in which $V_{dcxf} = \sqrt{2}V_{invxf}$, $V_{dcxn} = \sqrt{2}V_{invxn}$. In order to provide sufficient dc-link voltage for compensating load reactive and harmonic currents, the minimum dc-link

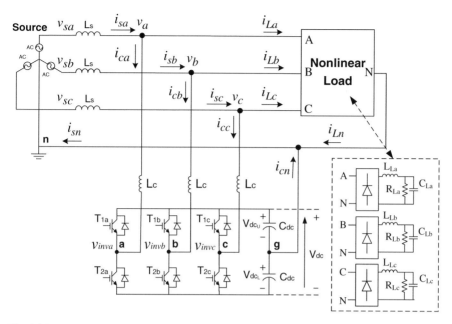

Fig. 3.83 Circuit configuration of the three-phase four-wire center-split APF

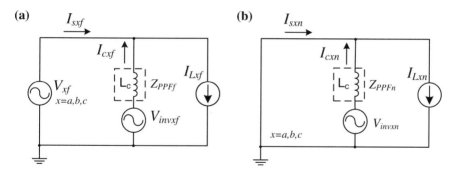

Fig. 3.84 APF single-phase equivalent circuit models at: **a** fundamental frequency and **b** harmonics frequency

voltage requirement (V_{dcx}) of the APF single-phase circuit model is deduced by considering the worst phase relation between each harmonic component, in which their corresponding peak voltages of the VSI at ac side are assumed to be superimposed.

$$V_{dcx} = \sqrt{|V_{dcxf}|^2 + \sum_{n=2}^{\infty} |V_{dcxn}|^2} \qquad (3.113)$$

From Fig. 3.11a, when the voltage load V_x is pure sinusoidal without harmonic components, $V_x = V_{xf}$. The inverter fundamental output voltage of the APF single-phase equivalent circuit model can be expressed as:

$$V_{invxf} = V_x + Z_{PPFf} \times I_{cxf} \qquad (3.114)$$

As the coupling part is a pure inductor L_c, $Z_{PPFf} = |X_{Lcf}|e^{j\phi f}$, where $\phi f = 90°$. When the APF operates in the ideal case, the fundamental compensating current I_{cxf} contains mainly reactive component I_{cxfq} with less active current component I_{cxfp}, therefore (3.114) can be rewritten as:

$$V_{invxf} \approx V_x + |\omega L_c||I_{cxfq}| \qquad \omega = 2\pi f \qquad (3.115)$$

From (3.115), it is clearly shown that V_{invxf} for APF must be larger than the load voltage V_x for reactive power compensation, no matter what the coupling inductor is. This also explains why the inverter part of the APF can only operate with a dc-link voltage higher than the peak of V_x value.

From Fig. 3.84b, the inverter harmonic output voltage of the APF single-phase equivalent circuit model V_{invxn} at each nth harmonic order can be expressed as:

$$V_{invxn} = |n\omega L_c||I_{cxn}| \qquad n = 2, 3 \ldots \infty \qquad (3.116)$$

where I_{cxn} is the nth order harmonic compensating current. When the APF is performing compensation, the absolute reactive and nth order harmonic compensating current should be equal to those of the loading, it yields:

$$|I_{cxfq}| = |I_{Lxfq}|, \qquad |I_{cxn}| = |I_{Lxn}| \qquad (3.117)$$

where I_{Lxfq} and I_{Lxn} are the reactive and nth order harmonic currents of the loading.

From (3.114) to (3.117), the inverter fundamental and each nth harmonic order output voltages of the APF single-phase equivalent circuit model (V_{invxf}, V_{invxn}) can be calculated. Then the minimum dc-link voltage requirement (V_{dcx}) for the APF single-phase circuit model can be found by (3.113).

By using the generalized single-phase p-q theory [32], the reactive power and current harmonics in each phase can be compensated independently, thus the final required minimum dc-link voltage for the three-phase four-wire center-split APF is the maximum among the calculated minimum value of each phase (V_{dcx}) as expressed in (3.118). Table 3.6 summarizes the minimum dc-link voltage deduction steps of the three-phase four-wire APF. (See Table 3.28)

$$V_{dc} = \max(2V_{dca}, 2V_{dcb}, 2V_{dcc}) \qquad (3.118)$$

Table 3.28 Minimum DC-link voltage deduction steps of the APF

1	Inverter fundamental output voltage:	
	$V_{invxf} = V_x + \lvert\omega L_c\rvert\lvert I_{cxfq}\rvert$	(3.115)
	where $\lvert I_{cxfq}\rvert = \lvert I_{Lxfq}\rvert, \quad \omega = 2\pi f$	(3.117)
2	Inverter nth harmonic order output voltage:	
	$V_{invxn} = \lvert n\omega L_c\rvert\lvert I_{cxn}\rvert$	(3.116)
	where	(3.117)
	$\lvert I_{cxn}\rvert = \lvert I_{Lxn}\rvert, \quad n = 2, 3\ldots\infty, \quad \omega = 2\pi f$	
3	Minimum dc-link voltage:	
	$V_{dc} = \max(2V_{dca}, 2V_{dcb}, 2V_{dcc})$	(3.118)
	where $V_{dcx} = \sqrt{\lvert V_{dcxf}\rvert^2 + \sum\limits_{n=2}^{\infty}\lvert V_{dcxn}\rvert^2}$	(3.113)
	$V_{dcxf} = \sqrt{2}V_{invxf}, \; V_{dcxn} = \sqrt{2}V_{invxn}$	

3.6.1.3   Simulation Verification for the APF Minimum DC-Link Voltage Analysis

Figure 3.85 shows the reactive and harmonic reference compensating current deduction and PWM control block diagram for the three-phase four-wire center-split APF, in which the three-phase v_x and i_{Lx} are needed in determining the reference compensating currents i_{cx}^* by the single-phase p-q theory [32]. The phase load voltage v_x and load current i_{Lx} (x = a, b, c) are transformed into their corresponding α-axis and β-axis quantities ($v_{x\alpha}, v_{x\beta}, i_{Lx\alpha}, i_{Lx\beta}$) in α-β coordinate by using a $\pi/2$ lag consideration. Then, the phase instantaneous active power p_{Lx} and reactive power q_{Lx} are calculated by $p_{Lx} = v_{x\alpha}i_{Lx\alpha} + v_{x\beta}i_{Lx\beta}$ and $q_{Lx} = v_{x\alpha}i_{Lx\alpha} - v_{x\beta}i_{Lx\beta}$. To compensate the reactive power and current harmonics generated by the load, i_{cx}^* for each phase can be calculated by $i_{cx}^* = \frac{1}{A}\left[-v_{x\alpha}\cdot\tilde{p}_{Lx} + v_{x\beta}\cdot q_{Lx}\right]$, where $A_x = v_{x\alpha}^2 + v_{x\beta}^2$. The term, oscillating part of instantaneous power, $\tilde{p}(\omega t)$ can easily be extracted from $p(\omega t)$ by using a low-pass filter (LPF) or high-pass filter (HPF). After the determination of the instantaneous current, i_{cx}^*, the compensating current error Δi_{cx} together with hysteresis band H will be sent to the current PWM control part. Then the PWM trigger signals for the switching devices of the VSI can then be generated. This control block diagram is applied in the APF system in order to generate the required compensating current, i_{cx}. In the following part, as the load harmonic current contents beyond the 9th order are small, for simplicity, the required minimum dc-link voltage calculation will be taken into account up to 9th harmonic order only.

In order to verify the minimum dc-link voltage analysis, representative simulation results of the three-phase four-wire center-split APF system in Fig. 3.83 are given as follows. The nonlinear loads are composed of three single-phase full bridge rectifiers. In order to simplify the verification, the dc-link is supported by external dc voltage source and the simulated three-phase loadings are approximately balanced. Table 3.29 lists the simulated system parameters for the APF and the current quality data of the loading. Simulation studies were carried out by using

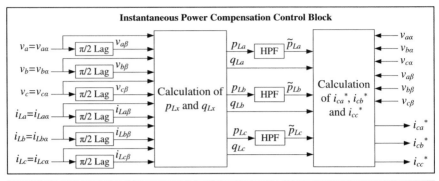

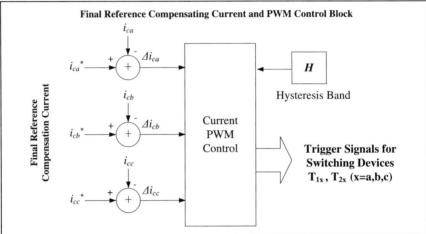

Fig. 3.85 Control block diagram for the three-phase four-wire center-split APF

Table 3.29 APF system parameters for simulations

System parameters		Physical values
System source-side	V_x, f	110V_{rms}, 50 Hz
	L_s	1 mH
APF	L_c, C_{dc}	30 mH, 10 mF
	V_{dcu}, V_{dcL}	180, 200, 220 V
Nonlinear rectifier load	R_{Lx}, L_{Lx}, C_{Lx}	26 Ω, 30 mH, 200 uF
Current quality data of LOAD	I_{Lx}	5.3A_{rms}
	DPF	0.835 (lagging)
	$\lvert I_{Lxfq} \rvert$	2.79A_{rms}
	$\lvert I_{Lx3} \rvert$	1.35A_{rms}
	$\lvert I_{Lx5} \rvert$	0.35A_{rms}
	$\lvert I_{Lx7} \rvert$	0.14A_{rms}
	$\lvert I_{Lx9} \rvert$	0.07A_{rms}

Table 3.30 Simulated results before and after APF compensation at different V_{dcu}, V_{dcL} levels

Simulation results	V_{dcu}, V_{dcL} (V)	i_{sx} (A)	DPF	THD_{isx} (%)	THD_{vx} (%)	i_{sn} (A)
Without comp.	–	5.30	0.835	30.0	1.82	4.09
APF	180	6.00	0.996	18.4	2.28	2.93
	200	5.15	1.000	12.5	2.30	1.60
	220	4.30	1.000	7.6	2.54	0.45

PSCAD/EMTDC. From Table 3.29, the required minimum dc-link voltage for the APF system can be calculated through (3.113)–(3.118), which is equal to $V_{dc} = 404.2\mathrm{V}(V_{dcu} = V_{dcL} = 202.1\ \mathrm{V})$.

Referring to Table 3.30, before compensation, the system current and system neutral current are $i_{sx} = 5.30$Arms and $i_{sx} = 4.09$Arms respectively. The displacement power factor (DPF) is 0.835, the total harmonic distortion of system current ($THDi_{sx}$) and load voltage ($THDv_x$) are 30.0 and 1.82 %, in which $THDi_{sx}$ does not satisfy the international standards ($THDi_{sx} < 15$ % for IEEE) [33].

Figure 3.86 and Table 3.30 show the simulated compensation results of APF with different dc-link voltage levels. When the dc-link voltage $V_{dcU} = V_{dcL} = 180$ V (<202.1 V), the APF operates at the rectifier mode, thus drawing active current. After compensation, the compensated $THDi_{sx} = 18.4$ % and i_{sn} is 2.93Arms, in which the compensated $THDi_{sx}$ does not satisfy the international standards. When the dc-link voltage increases to $V_{dcU} = V_{dcL} = 200$ V, since this value is closer to the required 202.1 V, the APF can obtain better compensating performances with $i_{sx} = 5.15$Arms, $THDi_{sx} = 12.5$ % and $i_{sn} = 1.60$Arms, in which the compensated $THDi_{sx}$ satisfied to the international standards. However, the APF is still operating at the rectifier mode due to insufficient dc-link voltage level. When the dc-link voltage increases to $V_{dcU} = V_{dcL} = 220$ V, which is higher than the minimum 202.1 V requirement, the APF can operate at both inverter and rectifier modes and achieve the best compensation performances with i_{sx} reduces to 4.30Arms, $THDi_{sx} = 7.6$ % and $i_{sn} = 0.45$Arms among three cases. Moreover, the compensation results as shown in Table 3.6 satisfy the international standards. From Table 3.30, the three cases can compensate the DPF from 0.835 to 0.995 or above. Moreover, the compensated $THDv_x$ satisfies the international standard [33].

Figure 3.86 and Table 3.30 have also verified the APF minimum dc-link voltage deduction. With different system voltage level, coupling inductor value and loading current contents, the APF requires different minimum dc-link voltage value for operation.

This section aims to investigate the minimum dc-link voltage design for three-phase four-wire center-split APF in reactive power and current harmonics compensation. First, its single-phase equivalent circuit models are built and proposed. Based on circuit models and current quality data of the loading, the minimum dc-link voltage for the three-phase four-wire APF is deduced and discussed. Simulation results of the APF are presented to verify this minimum dc-link voltage.

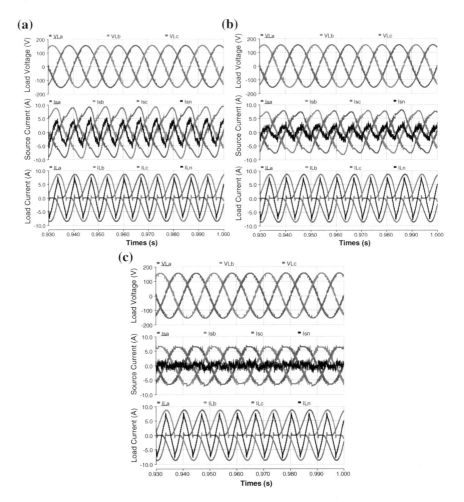

Fig. 3.86 APF simulation results: **a** V_{dcU}, V_{dcL} = 180 V, **b** V_{dcU}, V_{dcL} = 200 V and **c** V_{dcU}, V_{dcL} = 220 V

3.6.2 Design of Coupling Inductor

A coupling inductance is connected between the point of common coupling (PCC) and the VSI in an APF, which directly affects the compensating currents. Two factors are considered for selecting the value of the coupling inductor, which are the ability to suppress the current ripple and current tracking speed.

3.6.2.1   Suppress Current Ripple

When the period of the PWM control is small enough, the current passing through the coupling inductor varies linearly, as given in (3.119) and illustrated in Fig. 3.87

$$\Delta I^* = \frac{v_{inv} - v}{L_c} \cdot T_s \tag{3.119}$$

where v_{inv} is the average output of the inverter in a switching period, v is the voltage at the PCC, and T_s is the period of PWM.

When PWM is used to control VSI, the output voltage of the VSI always changes discontinuously. For example, the output voltage varies between 0 and V_{dc} for a two-level VSI. As shown in Fig. 3.87, the output voltage of the VSI varies between two neighboring levels to synthesize the reference voltage. The average output voltage in each period equals to v_{inv}, as expressed in (3.120).

$$v_{inv} \cdot T_s = V_{dc} \cdot DT_s \tag{3.120}$$

Equation (3.120) is for two-level VSI and it needs to be revised as (3.121) for a multi-level VSI [34]. v_{inv} is assumed to be in the range of V_{offset} to $V_{offset} + E$. E denotes voltage of each level, D is the duty ratio and V_{offset} is integer multiples of E.

$$v_{inv} \cdot T_s = (V_{offset} + V_{dc}) \cdot DT_s + (1 - D)T_s \cdot V_{offset} \tag{3.121}$$

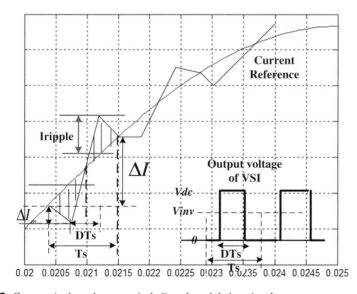

Fig. 3.87 Current ripple and symmetrical-aligned modulation signal

For simplicity, (3.120) for two-level inverter is used hereinafter. However, it can be proved that all the deduced results also work for multi-level inverters. The source voltage is assumed to be a constant in a period. The current passing through the coupling inductor varies like a saw tooth wave. Symmetrical-aligned modulation signal is selected since it gives the lowest current ripple [18]. The current variation in a period is shown in Fig. 3.87 and is expressed in (3.122).

$$\Delta I = \frac{-v}{L_c} \cdot \frac{1-D}{2} T_s + \frac{V_{dc} - v}{L_c} \cdot DT_s + \frac{-v}{L_c} \cdot \frac{1-D}{2} T_s \qquad (3.122)$$

According to (3.120), the ΔI^* in (3.119) and ΔI in (3.122) are the same. Practically, dead-time during the switching, delay in the control system and errors in the inductor value generate errors between ΔI^* and ΔI. They may not match exactly as shown in Fig. 3.87. All these factors are overlooked since they are not the main cause of the current ripple. The current ripple exists mainly due to the output voltage of the inverter changes discontinuously. As shown in Fig. 3.87, the error between the compensating current and its reference are two shadowed areas. Due to the symmetrical properties, the areas of the two shadowed error are the same and the total error area is calculated by (3.123),

$$\begin{aligned} S_{error} &= 2\left[\frac{1}{2}\frac{T_s}{2}\frac{\Delta I}{2} + \frac{1}{2}(\frac{T_s}{2} + \frac{DT_s}{2})\Delta I_m - \frac{1}{2}\frac{DT_s}{2}(\frac{\Delta I}{2} + \Delta I_m)\right] \\ &= \frac{1}{2}T_s(\frac{1-D}{2}\Delta I + \Delta I_m) \end{aligned} \qquad (3.123)$$

where ΔI_m is the change of the APF output current from 0 to $(1 - D)T_s/2$, and $\Delta I_m = \frac{v}{L_c}\frac{1-D}{2}T_s$. Combined with (3.120) and (3.122), (3.123) could be further revised as:

$$S_{error} = \frac{T_S^2}{4L_c}(DV_{dc} - D^2 V_{dc}) \qquad (3.124)$$

Since D is the duty ratio and varies from 0 to 1, S_{error} increases as V_{dc} increases. According to (3.125), dS_{error}/dD equals to zero when D = 1/2, i.e., the maximum value of S_{error} occurs when D = 1/2.

$$\frac{dS_{error}}{dD} = \frac{T_s^2}{4L_c}(V_{dc} - 2DV_{dc}) = 0 \qquad (3.125)$$

I_{ripple} is defined as the maximum variation of the output current of the APF from the reference current in a period. When D = 1/2, I_{ripple} is

$$I_{ripple} = \Delta I_m + \frac{(1-D)}{2}\Delta I = \frac{T_s V_{dc}}{8L_c} \qquad (3.126)$$

If the current ripple is set to be smaller than ΔI_r, (3.127) is obtained.

$$L_c \geq \frac{V_{dc}}{8 f_S \Delta I_r} \tag{3.127}$$

For multi-level inverter, the dc bus voltage V_{dc} is split to $(N-1)$ levels, and (3.127) is revised as (3.128).

$$L_c \geq \frac{V_{dc}}{8(N-1) f_s \Delta I_r} \tag{3.128}$$

In addition, if left-aligned or right-aligned modulation is used instead of symmetrical-aligned modulation, the corresponding current ripple is (3.129). The current ripple is doubled. Hence, the symmetrical-aligned modulation is better for suppressing current ripple and is suggested to be adopted in APFs. Finally, (3.128) is employed for handling coupling inductor values.

$$I_{ripple} = \frac{T_s V_{dc}}{4 L_c} \tag{3.129}$$

3.6.2.2   Current Tracking Speed

The relationship between the coupling inductor and the slope of the compensating current is:

$$\frac{di_c}{dt} = \frac{v_{inv} - v}{L_c} \tag{3.130}$$

where i_c is the compensating current of APF. When reactive current and current harmonics are compensated simultaneously by APF, the current slope is expressed as:

$$\frac{di_c}{dt} = \frac{di_{cfq}}{dt} + \frac{di_{cn}}{dt} \tag{3.131}$$

where $\frac{di_{cfq}}{dt} = \sqrt{2} I_{cfq} \omega \cos(\omega t - 90°) = \sqrt{2} I_{cfq} \omega \sin \omega t$

$$\frac{di_{cn}}{dt} = \frac{d\left(\sum_{n=2}^{\infty} \sqrt{2} I_{cn} \sin(n\omega t - \varphi_h)\right)}{dt} = \sum_{n=2}^{\infty} \sqrt{2} I_{cn} n\omega \cos(n\omega t - \varphi_h)$$

In order to track the compensating current, the right side of (3.130) should be larger than the maximum current slope. Hence, (3.132) is obtained.

$$L_c \leq \frac{v_{inv} - v}{\sqrt{2}I_{cfq}\omega + \sum\limits_{n=2}^{\infty} \sqrt{2}I_{cn}n\omega \cos(n\omega t - \varphi_h)} \tag{3.132}$$

It's difficult to predict the voltage difference at the instant when the maximum current changing rate is required. In order to calculate the boundary of the inductor value according to (3.132), complex measurement of the load currents are necessary to provide the required parameters. For simplicity, (3.133) is proposed to replace (3.132) in order to estimate the upper boundary of the coupling inductor of the APF.

$$L_c \leq \frac{\delta_v V_{dc}}{rI_c\omega} \tag{3.133}$$

It is suggested to select δ_v in the range of 0.1–0.3. I_c is the current rating of the APF and r is suggested to select the order of the most significant harmonic in the load currents. It is easy to estimate the r value without measuring the load currents. When the loads are mainly home or office appliances, which uses single-phase switch mode power supply, r should be three. If the load is a three-phase adjustable speed drive, r usually selects five. With the determination of L_c, the current tracking speed can be determined.

3.6.2.3 Method for Calculating Coupling Inductance

Combining the results of the previous two parts; the range of the coupling-inductance of an APF is expressed as (3.134). The detailed explanations for each parameter are given in Table 3.31. The lower boundary is totally determined by the APF settings. The last two parameters in Table 3.31 vary with non-linear loads to be compensated and are used for estimating the upper boundary.

$$\frac{V_{dc}}{8f_s(N-1)\Delta I_r} \leq L_c \leq \frac{\delta_v V_{dc}}{r\omega I_c} \tag{3.134}$$

Table 3.31 Parameters for calculating inductor value

Parameter	Explanation	Example
V_{dc}	DC bus voltage of the inverter	200 V
N	The level of the VSI	3
f_s	PWM frequency	5000
I_c	Current rating of the APF (RMS)	5 A
ω	Fundamental frequency	314
ΔI_r	Maximum current ripple. Typical value: 10 % Ic	0.5 A
δ_v	Typical value: 0.1–0.3	0.2
r	The order of the most significant harmonic	3

When the final value of the inductor is determined, a value close to the lower boundary should be selected because it could provide better current tracking speed with acceptable current ripples. A smaller inductor also reduces the cost. However, the final value should be tested by estimating the upper boundary according to (3.134). In some cases, the derived upper limit of the inductor value is smaller than the lower limit. It indicates conflicts exist in the requirement of current tracking speed and suppressing current ripple. Some approaches, such as increasing PWM frequency and adopting multi-level VSI, could be considered to reduce the lower boundary. Another alternative solution is to use LR, LC or LCL filter in the coupling circuits [35].

3.6.2.4 Simulation Results

Simulation models are built by using PSCAD/EMTDC. Parameters are listed in Table 3.31 and the RMS value of source voltage V is 55 V. The simulation model of an APF using three-level center-split VSI is first built. The L_c lower boundary is 5 mH by using the proposed method and the L_c upper boundary is 8.4 mH. Figure 3.88 shows the performance of the APF to track the compensating current reference for three different inductor values. When inductor value is 5 mH, the source currents are shown in Fig. 3.89, in which the APF begins compensation at 0.04 s.

3.6.3 Implementation of a Three-Phase Four-Wire APF Prototype

3.6.3.1 System Configurations of Three-Phase Four-Wire APFs

Prototypes of three-phase four-wire active power filters have been developed. The system configuration of the three-phase four-wire active power filter is shown in Fig. 3.90. If switch "K" is connected to the mid-point of the dc bus, three-leg center-split VSI is used. Otherwise, the switch "K" is connected to the fourth-leg of the VSI and a four-leg VSI is used.

The control system of the APP includes three main blocks. Descriptions about these blocks are provided in the following sections.

- Sampling circuits
- Digital control system
- Switching devices and their driving circuits

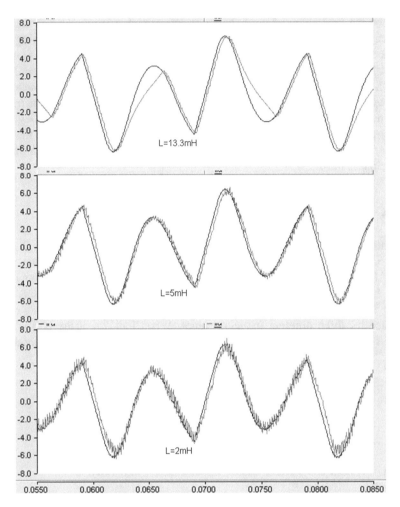

Fig. 3.88 Output current and reference current of the APF using three-level VSI

3.6.3.2   Sampling Circuits

The system voltages, load currents and injecting currents from the APF are measured by transducers. These signals are required to determine the compensating currents and reference voltages. The voltage and current transducers are selected according to the working frequency, work rating, the required accuracy, linearity and response time. For example, both current transformers (CT) and Hall Effect transducers can be used to measure the currents. CT is much cheaper especially for measuring high currents. However, it introduces phase shifting to the detected signal. For high frequency component in the current, the saturation effect of the CT also distorted the current waveform.

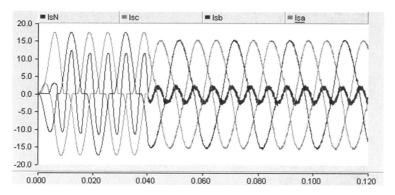

Fig. 3.89 Source current when $L_c = 5$ mH

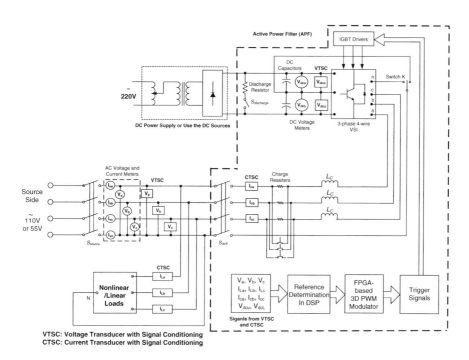

Fig. 3.90 System configuration of the APF prototypes

The prototype adopts transducers that are based on the Hall effects, which can transfer large electrical signals into small analog signals. The photo of the voltage and current transducer board is shown in Fig. 3.91. The output analog signals of the transducers are ac voltages, and its range is between −5 and 5 V.

The measured signals are finally sent to the A/D converter and converted to digital signals. If the output signals of the transducer do not satisfy the voltage

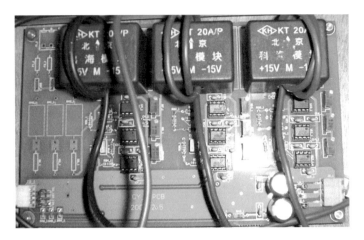

Fig. 3.91 The transducer boards

requirements of the analog to digital (A/D) converter, signal conditioning circuits need to be added. In the prototype of the APF, the signal conditioning circuits are implemented on the same board with the transducers. The schematic figure of the signal conditioning circuits is shown in Fig. 3.92.

The selected A/D converter can accept input analog signals in the range of 0–3.3 V. Therefore, the output voltage of the sampling circuits should be consistent with (0–3.3 V). In order to avoid the saturation when the input voltage approaches 3.3 V, the peak output voltage is changed to the range between 0.3 and 3.0 V. According to the schematic shown in Fig. 3.92, the signal conditioning equation is expressed as:

$$v_0 = \frac{R5}{R2} v_{in} + \frac{R5}{R3} \times 15 \qquad (3.135)$$

Set R5 = 2.2 kΩ and R3 = 20 kΩ, then $v_o = \frac{2.2k}{R2} v_{in} + 1.65$. The peak value of the output voltage of the transducers equals to $5\sqrt{2}$. R2 is 11 kΩ, so that the output

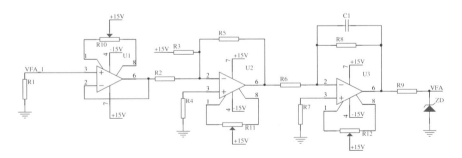

Fig. 3.92 The schematic figure of the signal conditioning circuits

voltage V_0 of the signal conditioning circuits is about 3.06 V. R10, R11, and R12 in Fig. 3.92 are adjustable resistor, which are used to adjust the zero point of the transducer output. R8 and C1 combine an active first order low-pass filter, which aims at filtering the high frequency noise signal. The final outputs of the signal conditioning circuits are sent to the A/D converter.

3.6.3.3 Digital Control System

The block diagram of the control system of the APF is given in Fig. 3.93. The reference voltage calculation is achieved based on TMS320F2407 controller from Texas Instrument. It is a low cost and high performance controller in real-time and motor/machine control. A sixteen-channel 10-bit A/D converter is embedded in the chosen DSP. There is a total of sixteen multiplexed analog inputs of 10-bit A/D converter core with built-in Sample and Hold (S/H). After conversion, the digital value of the selected channel is stored in the appropriate result register.

The reference voltages are written to the data buffers in 3D PW modulator. The DSP only implements the sampling and reference calculations. The FPGA-based 3D PW modulator generates the trigger signals based on the 3D direct PWM. The programs in DSP are mainly composed of two parts, as shown in Fig. 3.94.

Figure 3.94a is the main program, which initializes the variables and the registers to make preparation for the sampling. In addition, the control registers, embedded in FPGA is also initialized, so that the PWM period, VSI topology and dead-band time are set for the FPGA-based PW modulator. After all the initializations are finished, the main program stops in an infinite loop and waits for the interrupts.

Figure 3.94b is the interrupt service routine (ISR), and the corresponding flow chart is given in Fig. 3.94b. The interrupt is triggered in every sampling period.

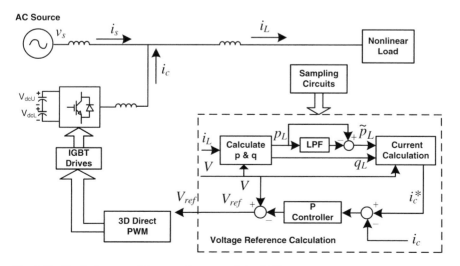

Fig. 3.93 Block diagram of the control system of the APF

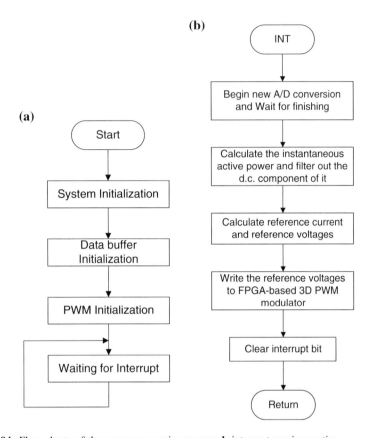

Fig. 3.94 Flow charts of the program **a** main program **b** interrupt service routine

The A/D conversion, reference signal calculations are all performed in this interrupt service routine. The obtained reference voltages are written to the registers inside the FPGA-based 3D PW modulator; then the corresponding trigger signals are generated by the generalized 3D PW modulator.

3.6.3.4 Switching Devices and Driving Circuits

The switching devices are selected for implementing the VSI. The introduction of the metal-oxide semiconductor (MOS) technology to the processing area of power semiconductors has created revolutionary device and application advantages. The presently available power switch category includes several device types including metal-oxide semiconductor field effect transistors (MOSFETs), gate turn off (GTO) thyristors, and insulated gate bipolar transistors (IGBTs).

The IGBT has the merit of low switching losses and require very little drive power at the gate. IGBT converters cover a power range up to several megawatt.

The switching frequencies range from a few kilohertz up to 20 kHz at derated power capability. The Mitsubishi third generation IGBT PM300DSA060 is selected in the two-level VSI prototype, which is a 300 A, 600 V, dual intelligent power module (IPM). The intelligent power module provides the user with the additional benefits of equipment miniaturization and reduced time to market as they include gate drive circuit and protection circuits [36, 37]. The appearance of the selected dual IPM is shown in Fig. 3.95a. It provides two series IGBT switches in one module, as shown in Fig. 3.95b [38]. This module can be used as one leg of a two-level VSI, which provides great convenience to the hardware implementation.

The schematic figure of the IPM drive board is shown in Fig. 3.96. The IPM drive board is made to control the switching state of IPM with the trigger signals from digital controller.

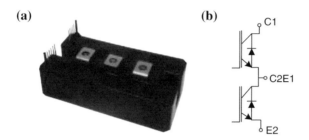

Fig. 3.95 IPM module **a** appearance, **b** structure of the dual IPM model

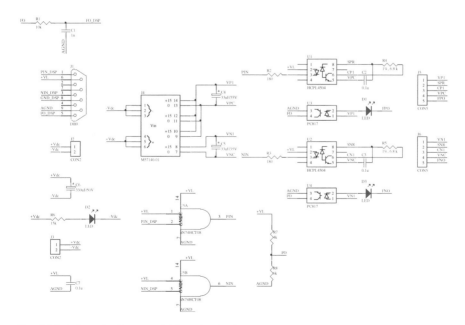

Fig. 3.96 Schematic diagram of driver circuits for IPM module

The followings are some considerations for the design of the IPM driver in consideration of FPGA I/O requirement.

- The SN74HCT08 is used to protect the I/O of FPGA. SN74HCT08 works as a buffer, since the current of I/O of FPGA is recommended to be kept within 1.67 mA per pin for 3.3 V-tolerance; while the typical working current of HCPL4504 is 16 mA.
- +VL, R7 and R8 are combined as resistor divider to pull up the signal "FO" and to limit the voltage of the pull-up signal "FO" within 3.3 V (not over the FPGA I/O voltage limit).
- The RC low-pass filter is adopted to filter the noise which could probably be on the fault signal. The selection of R1 should consider the current of FPGA I/O, "FO" pull-up potential, and RC low-pass filter's load effect.

3.6.4 Experimental Results

3.6.4.1 Three-Phase Four-Wire APF Using a Two-Level Center-Split VSI

The experimental results for the three-phase four-wire APF, in which a two-level three-leg center-split VSI is used, are given in this section. The system configuration of the APF is shown in Fig. 3.97, where the neutral wire is connected to the mid-point of the dc bus.

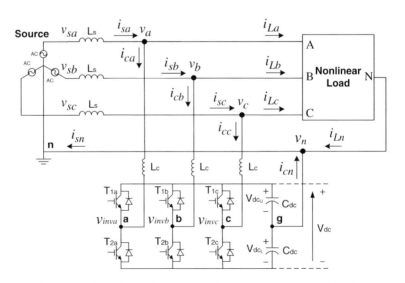

Fig. 3.97 Three-phase four-wire APF using a two-level three-leg center-split VSI

The generalized 3D direct PWM is applied to control the two-level three-leg center-split VSI. Figure 3.98 gives the source and neutral currents and its corresponding THD% before compensation. The RMS value of source currents is around 0.93 A and THD is about 32.1 %. The main harmonics are 3rd, 5th and 7th which are determined by the nonlinear loads. Figure 3.99 shows the source and the neutral currents and the corresponding THD after compensation. The THD of source current reduces to 7.4 % after compensation. Results indicate that the current harmonics are compensated and neutral current is reduced by the three-phase four-wire APF.

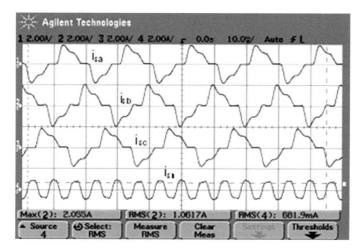

Fig. 3.98 Source and neutral currents before APF compensation

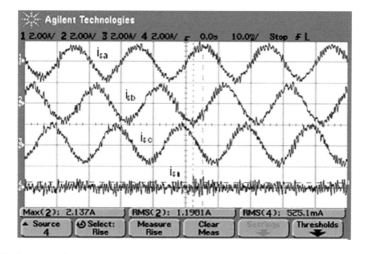

Fig. 3.99 Source and neutral currents after APF compensation

3.6.4.2 Three-Phase Four-Wire APF Using a Two-Level Four-Leg VSI

The system configuration of the APF is shown in Fig. 3.100, in which a two-level four-leg VSI is used. In this case, the neutral wire is connected to the fourth-leg of the voltage source inverter. The generalized 3D direct PWM is applied to control the four-leg VSI.

The system voltage, load currents and source currents after compensation are shown in Fig. 3.101. The THD of the load currents is about 30.2 %. The THD of the

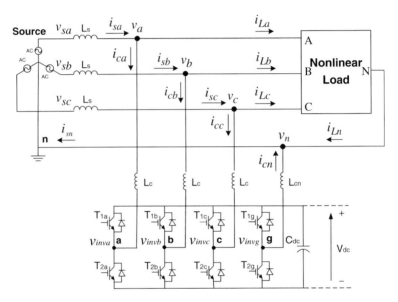

Fig. 3.100 Three-phase four-wire APF using a two-level four-leg VSI

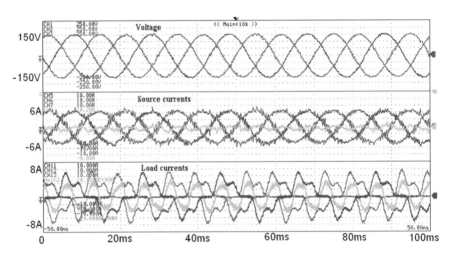

Fig. 3.101 Current compensation using a two-level four-leg VSI based APF

source currents after compensation is reduced to 7.1 %. The neutral current is also greatly reduced after compensation. Therefore, the validity of 3D direct PWM in controlling a two-level four-leg VSI in a three-phase four-wire APF is also proved.

3.6.4.3   Three-Phase Four-Wire APF Using a Three-Level Center-Split VSI

In this section, a three-level three-leg center-split VSI is used in the three-phase four-wire APF. The system configuration of the APF is shown in Fig. 3.102. The generalized 3D direct PWM is applied to control the three-level center-split VSI.

The load currents are shown in Fig. 3.103. The source currents after compensation are shown in Fig. 3.104. The corresponding THD values of the source currents and load currents are listed in Table 3.32. The RMS values of the neutral current before and after compensation are also given in Table 3.32. The results indicate that the current harmonics and neutral current are compensated.

The reactive power compensation performance is illustrated in Fig. 3.105. The corresponding power factor and displacement power factor are listed in Fig. 3.105.

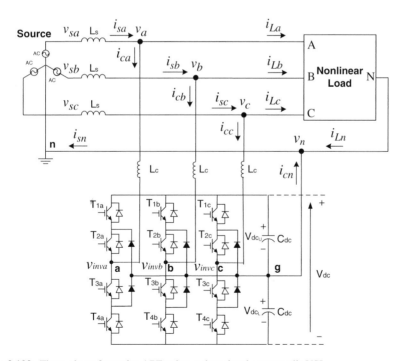

Fig. 3.102   Three-phase four-wire APF using a three-level center-split VSI

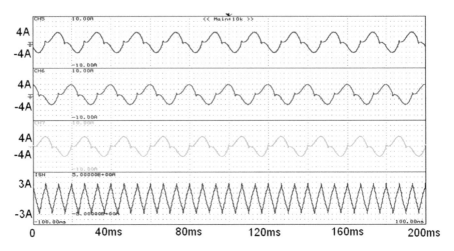

Fig. 3.103 Load currents

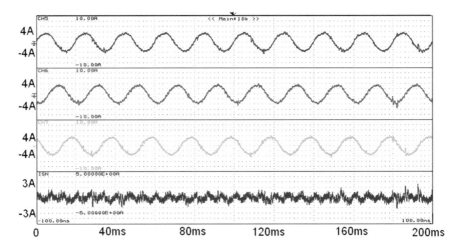

Fig. 3.104 Source current after APF compensation using a three-level center-split VSI

Table 3.32 Compensation results of APF when a three-level center-split VSI is used

	THD A (%)	THD B (%)	THD C (%)	Neutral (A)
Load currents	25.2	25.3	25.4	1.56
Source currents after compensation	7.5	6.5	6.2	0.53

(a)   **(b)**

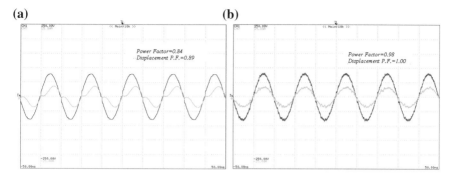

Fig. 3.105 Reactive power compensation: **a** voltage and load current, **b** voltage and source current after compensation

It can be found that the power factor approaches one after the reactive power is compensated by the APF. Therefore, the APF also can compensate reactive power if proper reference generation algorithms are used.

3.7 Summary

In this chapter, active power filters are discussed. Basically, an active power filter is constructed by an inductor connecting in series with a power electronics converter. The development of active power filters is discussed in the first section. However, the main purpose for this book is to introduce the three-phase four-wire power quality compensators. As a result, four-leg and three-leg center-split VSIs are mainly mentioned. Mathematical models of the four-leg and three-leg center-split VSIs as active power filters are given in the corresponding sections. This chapter can be summarized as follows:

(1) Four-leg and three-leg center-split VSI operated as three-phase four-wire power quality compensators are discussed. Its related voltage control reference signals corresponding to the required compensating currents as well as current control algorithms are mentioned.

(2) Based on the space basis, three-dimension space vectors can be deduced. Space vector allocation and their amplitudes of four-leg and three-leg center-split VSIs are compared and discussed.

(3) The corresponding hysteresis PWM, sign rectangular bar hysteresis PWM, cylindrical coordinate control, space vector modulations, direct PWM for control strategies are given.

(4) DC linked voltage control and comparison are discussed.

(5) Multi-level topology comparison and generalized control method for three-phase four-wire systems as active power filters are given.

References

1. M. takeda, K. Ikeda, Y. Tominaga, Harmonic current compensation with active filters, in *Conference Record of IEEE-IAS Annual Meeting*, pp. 808–815 (1987)
2. H. Kawahira, T. Nakamura, S. Nakazawa, M. Nomura, Active power filter, in *Conference Record of IPEC-Tokyo*, pp. 981–992 (1983)
3. S. Itoh, S. Fukuda, Basic characteristics of active power filter using current source converter. Ind. Appl. Soc. National Convention Rec. of IEE Jpn (1988)
4. L. Gyuyi, E.C. Strycula, Active ac power filter, in *Proceeding of 1976 IEEE/IAS Annual Meeting* (1976)
5. B. Singh, K. Al-Haddad, A. Chandra, A Reviews of active filters for power quality improvement. IEEE Trans. Indus. Electron. **46**(5), 960–971 (1999)
6. C.A. Quinn, N. Mohan, Active filtering of harmonic currents in three-phase, four-wire systems with three-phase, single-phase nonlinear loads, in *Proceedings IEEE APEC'92*, 1992, pp. 829–836
7. M.C. Wong, N.Y. Dai, J. Tang, Y.D. Han, Theoretical study of 3 dimensional hysteresis techniques. Proc. International Conf. Power Syst. Technol. (PowerCon) **1**, 560–564 (2002)
8. N. Y. Dai, S.U. Tai, M.C. Wong, Y.D. Han, Parallel power quality compensation in modern buildings, in *Proceedings of ICPESA*, Nov. 2004, pp. 54–60
9. N.Y. Dai, M.C. Wong, Y.D. Han, Application of a three-level NPC inverter as a 3-Phase 4-Wire power quality compensator by generalized 3DSVM. IEEE Trans. Power Electron. **21**, 440–449 (2006)
10. M.C. Wong, J. Tang, Y.D. Han, Cylindrical coordinate control of 3-dimensional PWM technique in 3-phase 4-wired tri-Level inverter. IEEE Trans. Power Electron. **18**(1), 208–220 (2003)
11. L. Malesani, P. Tomasin, PWM current control techniques of voltage source converters—a survey, in *Proceedings of IECON'93*, Nov. 1993, vol. 2, pp. 670–675
12. L.H. Hoang, L.A. Dessaint, An adaptive current control scheme for PWM synchronous Motor Drives: analysis and simulation. IEEE Trans. Power Electron. **4**(4), 486–495 (1989)
13. H. Agaki, Y. Kanazawa, A. Nabae, Instantaneous reactive power compensators comprising switching devices without energy storage components. IEEE Trans. Indus. Appl. **IA-20**(3), 625–630 (1984)
14. F.Z. Peng, G.W. Ott, D.J. Adams, Harmonic and reactive power compensation based on the generalized instantaneous reactive power theory for three-phase four-wire systems. IEEE Trans. Power Electron. **13**(6), 1174–1181 (1998)
15. B.J. Kang, C.M. Liaw, Robust hysteresis current-controlled PWM scheme with fixed switching frequency. IEE Proceedings—Electric Power Appl. **148**(6), 503–512 (2001)
16. M.C. Wong, N.Y. Dai, J. Tang, Y.D. Han, Theoretical study of 3 dimensional hysteresis PWM techniques, in *The 4th International Power Electronics and Motion Control Conference, 2004, IPEMC 2004*, vol. 3, pp. 1635–1640
17. H.W. Van Der Broeck, H.-C. Skudelny, G.V. Stanke, Analysis and realization of a pulsewidth modulator based on voltage space vectors. IEEE Trans. Indus. Appl. **24**(1), 142–150 (1988)
18. R. Zhang, V.H. Prasad, D. Boroyevich, F.C. Lee, Three-dimensional space vector modulation for four-leg voltage-source converters. IEEE Trans. Power Electron. **17**(3), 314–326 (2002)
19. J.H. Kim, S.K. Sul, A carrier-based PWM method for three-phase four-leg voltage source converters. IEEE. Trans. Power Electron. **19**, 66–75 (2004)
20. N.Y. Dai, M.C. Wong, F. Ng, Y.-D. Han, A FPGA-based generalized pulse width modulator for three-leg centre-split and four-leg voltage source inverters. IEEE Trans. Power Electron. **23**(3), 1472–1484 (2008)
21. M. Aredes, J. Hafner, K. Heumann, Three-phase four-wire shunt active filter control strategies. IEEE Trans. Power Electron. **12**(2), 311–318 (1997)
22. P. Verdelho, G.D. Marques, Four-wire current regulator PWM voltage converter. IEEE Trans. Indus. Electron **45**, 761–770 (1998)

23. P. Salmeron, J.C. Montano, J.R. Vazquez, J. Prieto, A. Perez, Compensation in nonsinusoidal, unbalanced three-phase four-wire systems with active power-line conditioner. IEEE Trans. Power Deliv. **19**(4), 1968–1974 (2004)

24. A. Nabae, I. Takahash, H. Akagi, A new neutral point clamped PWM inverter. IEEE Trans. Indus. Appl. **17**(5), 518–523 (1981)

25. D. Zhou, D.G. Round, Experimental comparisons of space vector neutral point balancing strategies for three-level topology. IEEE Trans. Power Electron. **16**(6) (2001)

26. M.-C. Wong, Z.-Y. Zhao, Y.-D. Han, L.-B. Zhao, Three-dimensional pulse-width modulation technique in three-level power inverters for three-phase four-wired system. IEEE Trans. Power Electron. **16**(3), 418–427 (2001)

27. N.-Y. Dai, M.-C. Wong, Y.-D. Han, 3 dimensional space vector modulation with dc voltage variation control in 3-leg centre-split power quality compensator. IEE J. Electric Power Appl. **151**(2), 198–204 (2004)

28. N.-Y. Dai, M.-C. Wong, Y.-D. Han, A 3-D generalized direct PWM algorithm for multilevel converters. IEEE Power Electron. Lett. **3**(3), 85–88 (2005)

29. N.-Y. Dai, M.-C. Wong, Y.-D. Han, Generalized pulse-width modulation control and controller for multi-level three-phase four-wire inverters. Chinese Invention Patent, Patent No.: 200710102959.6

30. N. Celanovic, D. Boroyevich, A fast space-vector modulation algorithm for multilevel three-phase converters. IEEE Trans. Indus. Appl. **37**(2), 637–641 (2001)

31. G. Garrara, S. Gardella, M. Marchesoni, R. Salutari, G. Sciutto, A new multilevel PWM method: a theoretical analysis. IEEE Trans. Power Electron. **7**, 497–505 (1992)

32. V. Khadkikar, A. Chandra, B.N. Singh, Generalized single-phase p-q theory for active power filtering: simulation and DSP-based experimental investigation. IET Power Electron. **2**, 67–78 (2009)

33. IEEE Recommended Practices and Requirements for Harmonic Control in Electrical Power Systems, 2014, IEEE Standard 519–2014

34. J.H. Kim, S.K. Sul, P.N Enjeti, A carrier-based PWM method with optimal switching sequence for a multi-level four-leg VSI, in *Proceedings of 40th IEEE Industry Application Conference*, vol.1, Oct. 2005, pp. 99–105

35. O. Vodyakho, C.C. Mi, Three-level Inverter-based shunt active power filter in three-phase three-wire and four-wire systems. IEEE Trans. Power Electron. **24**(5), 1350–1363 (2009)

36. J.D. Van Wyk, F.C. Lee, Power electronics technology at the dawn of the new millennium-status and future, in *Proceedings of IEEE Power Electronics Specialists Conference*, vol.1, June-July 1999, pp. 3–12

37. Featured projects technology and trend, Mitsubishi Electric., Japan, 1998

38. PM300DSA060: Flat-base type insulated package datasheet, Mitsubishi Electric., Japan, 1998

Chapter 4
Hybrid Active Power Filters

Abstract In this chapter, the hybrid active power filters are discussed. In Sect. 4.1, the development of hybrid active power filters (HAPFs) is initially presented. Then different hybrid active filter (HAPF) topologies have been compared and discussed in details in Sect. 4.2. Among them, LC-coupling HAPF (LC-HAPF) topology is being chosen for further investigation in this chapter because it can offer the lowest cost, size and weight, and has potential to provide dynamic reactive power compensation. Owning to the limitations or existing problems of the LC-HAPF as discussed in Sect. 4.3, this chapter aims to provide their corresponding solutions one by one. In Sect. 4.4, the harmonic resonances prevention, compensation capabilities and system robustness of the LC-HAPF are studied and investigated based on the deduced circuit model. Then different dc-link voltage control techniques for the LC-HAPF while performing dynamic reactive power compensation are explored and presented in Sect. 4.5. A novel adaptive dc-link voltage control technique for LC-HAPF in reducing switching loss and switching noise under reactive power compensation is proposed in Sect. 4.6. This adaptive control technique is also applied to active power filter (APF) for investigating switching loss and switching noise reduction. In Sect. 4.7, a minimum inverter capacity design for LC-HAPF in dynamic reactive power and current harmonics compensation is also studied and discussed. Finally, the design and performance of a LC-HAPF experimental system will be presented and discussed in Sect. 4.8, which can effectively compensate both dynamic reactive power and current harmonics in the three-phase four-wire distribution power systems.

4.1 Development of Hybrid Active Power Filters (HAPFs)

In order to improve the compensation characteristics of Passive Power Filters (PPFs) and to reduce the voltage and/or current ratings (costs) of the Active Power Filters (APFs), different hybrid active power filter (HAPF) topologies composed of active and passive filters in series and/or parallel have been examined in past researches [1–26]. They can be classified into three general types: HAPF topology

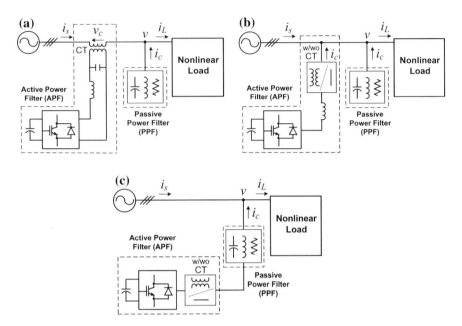

Fig. 4.1 Three general types of HAPF topology: **a** HAPF topology 1—series APF and shunt PPF, **b** HAPF topology 2—shunt APF and shunt PPF and **c** HAPF topology 3—APF in series with shunt PPF

1—series APF and shunt PPF [1–8], HAPF topology 2—shunt APF and shunt PPF [1, 2, 9–13], and HAPF topology 3—APF in series with shunt PPF [2, 13, 17–26] as shown in Fig. 4.1.

4.1.1 HAPF Topology 1—Series APF and Shunt PPF [1–8]

The HAPF topology 1—series APF and shunt PPF circuit configuration is shown in Fig. 4.1a. Figure 4.2 shows this HAPF topology in three-phase three-wire power system presented by Peng et al. in 1988 [4] and three-phase four-wire power system presented by Salmeron and Litran in 2010 [8] respectively. The APF is connected in series with the distribution power system through filtering inductor, capacitor (LC) and coupling transformer (CT) while the shunt PPF can be a tuned LC filter, high pass filter or any combination of them.

Under this HAPF topology, the APF acts as a harmonic isolator between the system source and load sides by forcing the load harmonic current flowing into the shunt PPF. At the fundamental frequency, the shunt PPF shows high impedance while the series APF shows low impedance. On the contrary, the shunt PPF shows low impedance while the series APF shows high impedance at harmonic frequency.

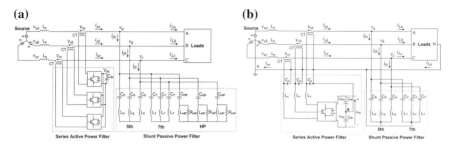

Fig. 4.2 HAPF topology 1—series APF and shunt PPF presented by: **a** Peng et al. in 1988 [4] and **b** Salmeron and Litran in 2010 [8]

This HAPF topology aims to reduce the current rating of the APF. Moreover, it can prevent the series and parallel resonance phenomena.

However, the series APF requires adequate protection in case of malfunction of the APF because it is in series with the distribution power system. Moreover, as the series APF is connected to the power system through CT, its CT is capable to withstand full load current, thus increasing the system cost, size and loss.

4.1.2 HAPF Topology 2—Shunt APF and Shunt PPF [1, 2, 9–13]

The HAPF topology 2—shunt APF and shunt PPF circuit configuration is shown in Fig. 4.1b. Figure 4.3 shows this HAPF topology with CT in three-phase three-wire power systems presented by Khositkasame and Sangwongwanich in 1997 [9] and Corasaniti et al. in 2009 [12], and without CT in three-phase four-wire power systems presented by Chiang et al. in 2005 [11] respectively. The shunt APF can be connected to the distribution power system through coupling (L) with or without CT while the shunt PPF can be a tuned LC filter, high pass filter or any combination of them.

Under this HAPF configuration, the PPF acts as main compensator and the APF is used to compensate the remaining current harmonic contents which have been filtered by the PPF, so as to improve the system filtering performance. Thus, this HAPF topology aims to reduce the current rating of the APF. In addition, the advantages of this topology are: the shunt APF is applicable if the shunt PPF already exists and the reactive power is controllable. Moreover, it can prevent the parallel resonance phenomenon.

However, the APF cannot change either the voltage across the PPF or the current through the PPF. Therefore, large circulating current will be generated by the PPF if the PPF impedance is low at the voltage distortion frequency. To prevent the harmonic current flows to the supply, the required APF current rating will still be high. If the APF is directly connected to the distribution power system without CT, its required voltage rating is high too. By adding CT, its voltage rating can be

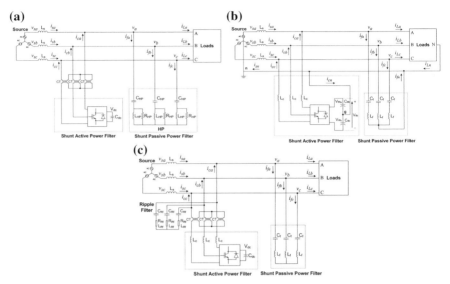

Fig. 4.3 HAPF topology 2—shunt APF and shunt PPF presented by: **a** Khositkasame and Sangwongwanich in 1997 [9], **b** Chiang et al. in 2005 [11] and **c** Corasaniti et al. in 2009 [12]

reduced; however, using CT will increase the system cost, size and loss. By choosing the PPF as a high pass filter, this HAPF topology can refrain from obtaining low PPF impedance at the voltage harmonic frequencies. However, high pass filters (with resistance) increase the filter losses and reduce the filtering effectiveness at the tuned frequency.

4.1.3 HAPF Topology 3—APF in Series with Shunt PPF [2, 3, 17–26]

The HAPF topology 3—APF in series with shunt PPF circuit configuration is shown in Fig. 4.1c. Figure 4.4 shows this HAPF topology with CT in three-phase three-wire power system presented by Fujita and Akagi in 1991 [3] and Rivas et al. in 2003 [17], and without CT in three-phase three-wire power system presented by Srianthumrong and Akagi in 2003 [20] respectively. The APF and PPF are connected in series with or without CT, and then this HAPF is shunt to the distribution power system. The PPF can be a tuned LC filter or high pass filter or any combination of them.

Under this HAPF configuration, the APF aims to change the impedance of the PPF so that the PPF has a nearly zero impedance to load-side current harmonics and infinite impedance to system-side voltage harmonics, so as to improve the compensation characteristics of the PPF. When a PPF and an APF are connected in series, the fundamental system voltage mainly drops on the capacitor of the PPF, but not that of the APF. Thus, this HAPF topology aims to reduce the voltage rating of the APF. Moreover, it can prevent the series and parallel resonance phenomena.

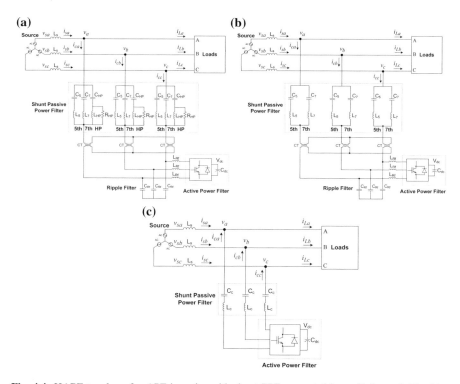

Fig. 4.4 HAPF topology 3—APF in series with shunt PPF presented by: **a** Fujita and Akagi in 1991 [3], **b** Rivas et al. in 2003 [17] and **c** Srianthumrong and Akagi in 2003 [20]

4.2 Comparison Among Three General HAPF Topologies

Table 4.1 shows the characteristics and comparisons among these three general HAPF topologies. They aim to perform current harmonic compensation only and can only inject a fixed amount of reactive power which is provided by its passive part. In practical case, the load-side reactive power consumption varies from time to time, they cannot perform satisfactory dynamic reactive power compensation. As shown in Table 4.1, HAPF topology 3 can be considered as the best one among the three general topologies because (i) it can effectively prevent series and parallel resonances, (ii) it has potential to generate less switching noise as it requires a low dc-link voltage and (iii) it does not contain large circulating current problem and can easily bypass the system in case of malfunction of APF or PPF.

In addition, the comparisons among the HAPF topology 3 [3, 17, 20] as shown in Fig. 4.4 are summarized in Table 4.2. From Table 4.2 and Fig. 4.4c seems to be an appropriated and cost-effective topology because it offers the lowest initial cost, size and weight. Moreover, it has the potential to provide the best dynamic reactive power compensation capability. Its coupling LC functions not only as a harmonic

Table 4.1 Characteristics and comparisons among three general HAPF topologies

	HAPF topology 1 [1–8]	HAPF topology 2 [1, 2, 9–13]	HAPF topology 3 [2, 3, 17–26]
Function of APF	Harmonic isolator	Harmonic compensator	To improve the filtering characteristics of PPF
Series and parallel resonance preventing capability	Effective on both	Effective on parallel resonance	Effective on both
Advantages	• Current capacity reduction of APF • Applicable even PPF already exists	• Current capacity reduction of APF • Applicable even PPF already exists	• Voltage capacity reduction of APF
Disadvantages	• Special isolation and protection during series APF malfunction • Extra loss due to CT	• Large circulating current generated during low PPF impedance at harmonic voltage • Extra loss—with CT	• Need modification if PPF already exists • Extra loss—with CT
Cost	High due to CT	High—with CT Lower—cwithout CT	High—with CT Lower—without CT
Size and weight	Bulky due to CT	Bulky—with CT Lighter—without CT	Bulky—with CT Lighter—without CT
Harmonic current compensation	Mainly support	Mainly support	Mainly support
Dynamic reactive power compensation	Possible to support	Possible to support	Possible to support

Table 4.2 Comparisons among HAPF topology 3 presented by: Fujita and Akagi in 1991 [3], Rivas et al. in 2003 [17] and Srianthumrong and Akagi in 2003 [20]

	Figure 4.4a	Figure 4.4b	Figure 4.4c
	Fujita and Akagi in 1991 [3]	Rivas et al. in 2003 [17]	Srianthumrong and Akagi in 2003 [20]
Cost	Highest	High	Lowest
Size and weight	Bulky	Bulky	Lightest
Coupling transformer	3	3	0
AC inductor	12	9	3
AC capacitor	12	9	3
Resistor	3	0	0
Harmonic current compensation	Best	Good	Medium
Dynamic reactive power compensation	N/A	N/A	N/A

filter but also as a switching ripple filter. In this Chapter, therefore, this LC coupling HAPF (LC-HAPF) topology is focused under three-phase distribution power system applications.

4.3   Existing Problems and Operation Principles of Conventional LC-HAPF

The LC-HAPFs are usually proposed in three-phase three-wire power systems [19–26]. When the LC-HAPF is applied to three-phase four-wire power systems, either center-split voltage source inverter (VSI) or four-leg VSI can be used. Compared with the four-leg VSI configuration, with just a slight increase in the operating dc-link voltage and an additional low cost dc capacitor for the three-leg center-split VSI, the center-split VSI configuration can save two costly power electronic switches and driver circuits [22]. Thus, the three-leg center-split VSI structure is chosen for the three-phase four-wire LC-HAPF. Figure 4.5 illustrates the system configuration of a three-phase four-wire center-split LC-HAPF, where the subscript 'x' denotes phase x = a, b, c. v_{sx} is the system voltage, v_x is the load voltage, Ls is the system inductance. i_{sx}, i_{Lx} and i_{cx} are the system, load and inverter compensating currents for each phase. C_{c1} and L_{c1} are the coupling capacitor and inductor of the LC-HAPF. C_{dc1}, V_{dc1U} and V_{dc1L} are the dc capacitor, upper and lower dc capacitor voltages of the LC-HAPF with $V_{dc1U} = V_{dc1L} = 0.5V_{dc1}$. The dc-link midpoint is assumed to be ground reference (g). From Fig. 4.5, the inverter line-to-ground voltages $v_{invlx\text{-}g}$ will be equal to the inverter line-to-neutral voltages $v_{invlx\text{-}n}$ because the neutral point n is connected to the dc-link midpoint g. The load can be a nonlinear load, a linear load or their combination. The nonlinear loads are composed of three single-phase diode rectifiers, which act as harmonic producing loads in this chapter.

The PPF part of the LC-HAPF is composed of a LC filter, in which the coupling C_{c1} and L_{c1} are designed based on the average fundamental reactive power consumption and the n_1th harmonic current order of the loading. The reactance of C_{c1} and L_{c1} can be expressed as:

$$X_{Cc1} = \frac{V_x^2}{|\bar{Q}_{Lxf}|} + X_{Lc1}, \quad X_{Lc1} = \frac{1}{n_1^2} X_{Cc1} \tag{4.1}$$

where V_x is the RMS load voltage, $\bar{Q}_{Lxf}$ is the phase average fundamental reactive power consumption of the loading. From (4.1), C_{c1} can be expressed as:

$$C_{c1} = \left(\frac{n_1^2 - 1}{n_1^2} \right) \frac{|\bar{Q}_{Lxf}|}{2\pi f \cdot V_x^2} \tag{4.2}$$

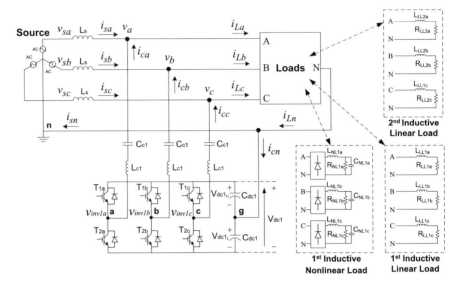

Fig. 4.5 System configuration of a three-phase four-wire center-split LC-HAPF

Then the coupling L_{c1} can be expressed as:

$$L_{c1} = \frac{1}{(n_1 \cdot 2\pi f)^2 \cdot C_{c1}} \tag{4.3}$$

where n_1 is the tuned harmonic order. This L_{c1} can be used to smooth the inverter output current ripple.

4.3.1 Existing Problems of Conventional LC-HAPF [19–26]

The existing problems of conventional LC-HAPF are listed below:

(1) Seldom overall study on its resonance prevention capability, filtering performance and system robustness.

(2) Its active VSI part is solely responsible for current harmonic compensation and it can only inject a fixed amount of reactive power which is provided by its coupling LC, thus the conventional LC-HAPF is a current harmonic-oriented design compensating system. As a result, it cannot perform satisfactory dynamic reactive power compensation because the load-side reactive power varies from time to time.

(3) LC-HAPF and other HAPF topologies are all operating at a fixed dc-link voltage level. Without adding extra soft-switching hardware components, they cannot reduce its switching loss and switching noise during operation.

(4) Its minimum inverter capacity requirement for reactive power and current harmonic compensation has not been mathematically deduced.

In this chapter, the corresponding solutions for the existing problems of the conventional LC-HAPF will be presented and discussed.

4.3.2   LC-HAPF Reference Compensating Current Determination Based on Single-Phase Instantaneous P-Q Theory

The compensation principle of the LC-HAPF is to generate a compensating current to the utility, so that it corrects the power factor close to unity and cancels harmonic currents. The reference compensating current calculation for the LC-HAPF can be achieved by the well-known instantaneous three-phase p-q theory proposed by Akagi et al. [27, 28] or the generalized single-phase p-q theory proposed by Khadkikar et al. [29].

The instantaneous p-q theory was originally developed for three-phase three-wire and four-wire systems [27, 28]. Recently, the p-q theory was developed and expanded for single-phase systems [29]. In addition, this single-phase p-q theory can also be well applied to three-phase or even multiphase systems. The single-phase p-q theory is based on an instantaneous $\pi/2$ lag or $\pi/2$ lead of voltage and current to define the original system as a pseudo two-phase system. Thus, the overall system can then be easily represented in α-β coordinates. The phase load voltage v_x and load current i_{Lx} (x = a, b, c) are considered as quantities on the α-axis, whereas β-axis quantities are obtained by a $\pi/2$ lag or $\pi/2$ lead of the phase load voltage and current [29]. The phase load voltage v_x and current i_{Lx} representation in α-β coordinates with a $\pi/2$ lag can be expressed as:

$$\begin{bmatrix} v_{x\alpha} \\ v_{x\beta} \end{bmatrix} = \begin{bmatrix} v_x \\ v_x e^{-j\pi/2} \end{bmatrix} \tag{4.4}$$

$$\begin{bmatrix} i_{Lx\alpha} \\ i_{Lx\beta} \end{bmatrix} = \begin{bmatrix} i_{Lx} \\ i_{Lx} e^{-j\pi/2} \end{bmatrix}, \tag{4.5}$$

where the subscript x = a, b, c. The phase instantaneous active power p_{Lx} and reactive power q_{Lx} can be expressed as:

$$\begin{bmatrix} p_{Lx} \\ q_{Lx} \end{bmatrix} = \begin{bmatrix} v_{x\alpha} & v_{x\beta} \\ -v_{x\beta} & -v_{x\alpha} \end{bmatrix} \cdot \begin{bmatrix} i_{Lx\alpha} \\ i_{Lx\beta} \end{bmatrix} \tag{4.6}$$

The p_{Lx} and q_{Lx} can also be expressed as:

$$p_{Lx} = \bar{p}_{Lx} + \tilde{p}_{Lx} \tag{4.7}$$

$$q_{Lx} = \bar{q}_{Lx} + \tilde{q}_{Lx}, \tag{4.8}$$

where $\bar{p}_{Lx}$ and $\bar{q}_{Lx}$ represent the DC components responsible for instantaneous fundamental active and reactive powers respectively, whereas $\tilde{p}_{Lx}$ and $\tilde{q}_{Lx}$ represent the AC components responsible for harmonic power.

To compensate the reactive power and current harmonics generated by the load, the reference compensating current i_{cx_q} for each phase of the LC-HAPF can be calculated by:

$$i_{cx\_q} = \frac{1}{A_x} \left[-v_{x\alpha} \cdot \tilde{p}_{Lx} + v_{x\beta} \cdot q_{Lx} \right] \tag{4.9}$$

where $A_x = v_{x\alpha}^2 + v_{x\beta}^2$. The term $\tilde{p}_{Lx}$ in (4.9) can easily be extracted from p_{Lx} by using a low-pass filter (LPF) or high-pass filter (HPF).

4.3.3   LC-HAPF Reactive and Harmonic Reference Compensating Current Determination and PWM Control Block Diagram

Figure 4.6 shows the reactive and harmonic reference compensating current deduction and PWM control block diagram for the three-phase four-wire LC-HAPF without dc-link voltage control, in which the dc link voltage control will be discussed in Sect. 4.5. After the instantaneous reference compensating current i_{cx_q} is determined, the final reference compensating current can be $i_{cx}^* = i_{cx\_q}$ without the dc control. Then the reference compensating current i_{cx}^* and the compensating current i_{cx} will be sent to the current PWM control part and the PWM trigger signals for the switching devices of the VSI can then be generated. And this control block diagram will be applied for LC-HAPF in order to generate the required inverter compensating currents.

In the following Sect. 4.4, the details of the LC-HAPF resonance prevention capability, filtering performance and system robustness will be studied and discussed. After that, the dynamic reactive power compensation and dc-link voltage control consideration issue for LC-HAPF will be presented and discussed in Sect. 4.5. Then an adaptive dc-link voltage control scheme for LC-HAPF dynamic reactive power compensation will be proposed in Sect. 4.6, so that the switching loss and switching noise can be reduced without adding extra soft-switching hardware components. In Sect. 4.7, the minimum inverter capacity design for the LC-HAPF will be deduced and analyzed. Then, the design and performance of a

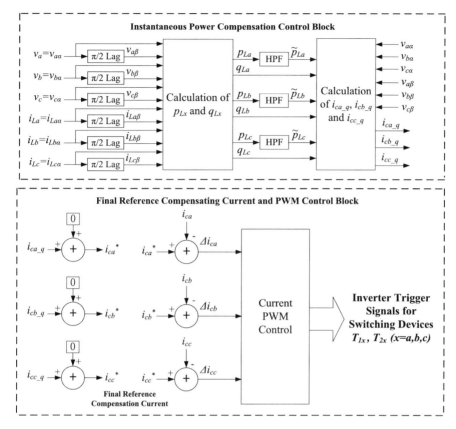

Fig. 4.6 Reactive and harmonic reference compensating current deduction and PWM control block diagram for the three-phase four-wire LC-HAPF

220 V/10 kVA three-phase four-wire LC-HAPF experimental prototype and its corresponding experimental results will be provided in Sect. 4.8. Finally, the summary for this chapter will be presented in Sect. 4.9.

4.4   Analysis of Three-Phase Four-Wire LC-HAPF Compensating Performances

In this section, the steady-state compensating performances for the LC-HAPF will be discussed, analyzed and compared with its pure passive power filter (PPF) part by four evaluation indexes. They are capabilities: (1) to prevent the parallel resonance between the PPF and the impedance of the power system, (2) to prevent the series resonance of the PPF, (3) to improve the filtering performance of the PPF and (4) to enhance the system robustness [30]. First of all, a single-phase harmonic equivalent circuit model of the three-phase four-wire LC-HAPF as shown in Fig. 4.5 will be deduced and built. Based on the harmonic circuit model, the steady-state compensating performances for both pure PPF part and LC-HAPF will be presented and discussed. Then their corresponding simulation results will be given to verify all the deduced and analyzed results. In the following, all the analyses are based on sufficient dc-link voltage assumption.

4.4.1 LC-HAPF Single-Phase Harmonic Circuit Model

From Figs. 4.5 and 4.7 shows the LC-HAPF single-phase harmonic equivalent circuit model, in which the nonlinear load and the active inverter part are modeled as a harmonic current source I_{Lxh} and voltage source V_{invxlh}. The subscript 'x' denotes phase a, b, c and the subscript 'h' denotes the harmonic component. V_{sxh} and V_{xh} are the system and load harmonic voltages, I_{sxh}, I_{Lxh} and I_{cxh} are the system, load and inverter compensating harmonic currents. Z_{sh} and Z_{PPFh} are the harmonic impedance of the power system and PPF respectively.

Provided that the inverter of the LC-HAPF is controlled by hysteresis PWM with hysteresis error band H ≈ 0, the inverter can be modeled as a current control voltage source. The i_{cx} direction is given in Fig. 4.5,

When $\left(i_{cxh}^{*} - i_{cxh}\right) \geq H$, i.e. $\left(i_{cxh}^{*} - i_{cxh}\right) \geq 0$, the inverter harmonic voltage can be expressed as:

$$v_{invxlh} = 0.5V_{dc1} = K_1 \cdot \left(i_{cxh}^{*} - i_{cxh}\right), \quad K_1 > 0 \tag{4.10}$$

When $\left(i_{cxh}^{*} - i_{cxh}\right) < -H$, i.e. $\left(i_{cxh}^{*} - i_{cxh}\right) < 0$,

$$v_{invxlh} = -0.5V_{dc1} = K_2 \cdot \left(i_{cxh}^{*} - i_{cxh}\right), \quad K_2 > 0 \tag{4.11}$$

Fig. 4.7 LC-HAPF single-phase harmonic circuit model

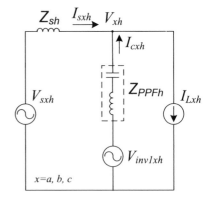

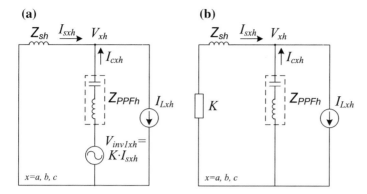

Fig. 4.8 LC-HAPF single-phase harmonic circuit model: **a** $V_{sxh} = 0$ and I_{Lxh} is considered only and **b** a harmonic equivalent circuit model due to I_{Lxh} only

where i^*_{cxh} represents the reference compensating harmonic current while i_{cxh} represents the actual compensating harmonic current. As K_1 and K_2 are both in positive, v_{invxlh} can be expressed in a general form as:

$$v_{invxlh} = K_1 \cdot \left(i^*_{cxh} - i_{cxh} \right), \quad K > 0 \tag{4.12}$$

In ideal case, i^*_{cxh} should be equal to the load harmonic current i_{Lxh}, that is $i^*_{cxh} = i_{Lxh}$. As $i_{sxh} + i_{cxh} = i_{Lxh}$, the inverter harmonic voltage can be expressed as:

$$v_{invxlh} = K_1 \cdot i_{sxh}, \quad K > 0 \tag{4.13}$$

From (4.13), Figs. 4.8 and 4.9 show the LC-HAPF single-phase harmonic circuit models due to load harmonic current i_{Lxh} or system harmonic voltage v_{sxh} components.

4.4.1.1   LC-HAPF Single-Phase Harmonic Circuit Model Due to I$_{Lxh}$ Only

When the system voltage contains no harmonic component, that is $V_{sxh} = 0$, the LC-HAPF single-phase harmonic circuit model due to load harmonic current I_{Lxh} is shown in Fig. 4.8a. From Fig. 4.8a,

$$I_{sxh} + I_{cxh} = I_{Lxh} \tag{4.14}$$

$$K \cdot I_{sxh} = I_{cxh} Z_{PPFh} - I_{sxh} Z_{sh} \tag{4.15}$$

From Eqs. (4.14) and (4.15), the harmonic currents flow into the power system I_{sxh} and the LC-HAPF I_{cxh} due to I_{Lxh} only can be expressed as:

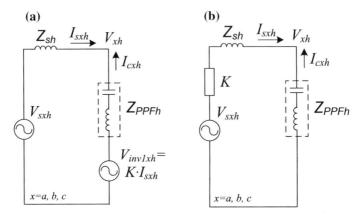

Fig. 4.9 LC-HAPF single-phase harmonic circuit model: **a** $I_{Lxh} = 0$ and V_{sxh} is considered only, **b** a harmonic equivalent circuit model due to V_{sxh} only

$$K_{sxh\_i} = \frac{I_{sxh}}{I_{Lxh}} = \frac{Z_{PPFh}}{K + Z_{sh} + Z_{PPFh}} \qquad (4.16)$$

$$K_{cxh\_i} = \frac{I_{cxh}}{I_{Lxh}} = \frac{K + Z_{sh}}{K + Z_{sh} + Z_{PPFh}} \qquad (4.17)$$

In a perfect LC-HAPF compensation, $K_{sxh\_i} = 0$ and $K_{cxh\_i} = l$ should be achieved so that the load harmonic current flows into the LC-HAPF ($I_{cxh} = I_{Lxh}$) and the power system does not contain any harmonic current ($I_{sxh} = 0$). In order to achieve this objective, K should be very large. Figure 4.8b shows a harmonic equivalent circuit model due to I_{Lxh} only. From Fig. 4.8b, it seems that the inverter is equivalent as a large harmonic impedance which is connected in series with the harmonic impedance of the power system Z_{sh}, so that it blocks the harmonic current flowing into the power system and forces it to flow into the LC-HAPF as deduced by (4.16) and (4.17). Therefore, the power system harmonic current I_{sxh} can be significantly reduced if K is significantly large.

4.4.1.2 LC-HAPF Single-Phase Harmonic Circuit Model Due to V_{sxh} Only

When the load current contains no harmonic components, that is $I_{Lxh} = 0$, the LC-HAPF single-phase harmonic circuit model due to system harmonic voltage V_{sxh} is shown in Fig. 4.9a. From Fig. 4.9a,

$$I_{sxh} = -I_{cxh} \qquad (4.18)$$

$$K \cdot I_{sxh} = I_{cxh}(Z_{PPFh} + Z_{sh}) + V_{sxh} \tag{4.19}$$

From (4.18) and (4.19), the harmonic currents flow into the power system I_{sxh} and the LC-HAPF I_{cxh} due to V_{sxh} only can be expressed as:

$$K_{sxh\_v} = \frac{I_{sxh}}{V_{sxh}} = \frac{1}{K + Z_{sh} + Z_{PPFh}} \tag{4.20}$$

$$K_{cxh\_v} = \frac{I_{cxh}}{V_{sxh}} = \frac{1}{K + Z_{sh} + Z_{PPFh}} \tag{4.21}$$

To minimize the effect of V_{sxh} to the harmonic currents in the power system and the LC-HAPF, $K_{sxh\_v} = 0$ and $K_{cxh\_v} = 0$ should be achieved. In order to achieve this objective, K should be very large. Figure 4.9b shows a harmonic equivalent circuit model due to V_{sxh} only. It seems that the inverter is equivalent to large harmonic impedance which is connected in series with the harmonic impedance of the power system Z_{sh}, so that the effect of V_{sxh} to the harmonic currents in the power system and the LC-HAPF can be significantly reduced if K is significantly large.

From (4.16), (4.17), (4.20) and (4.21), with superposition theorem, the power system harmonic current I_{sxh} and the LC-HAPF compensating harmonic current I_{cxh} due to V_{sxh} and I_{Lxh} can be expressed as:

$$I_{sxh\_v} = \frac{1}{K + Z_{sh} + Z_{PPFh}} V_{sxh} + \frac{Z_{PPFh}}{K + Z_{sh} + Z_{PPFh}} I_{Lxh} \tag{4.22}$$

$$I_{cxh\_v} = \frac{1}{K + Z_{sh} + Z_{PPFh}} V_{sxh} + \frac{K + Z_{sh}}{K + Z_{sh} + Z_{PPFh}} I_{Lxh} \tag{4.23}$$

When only pure PPF part is used, $K = 0$ in (4.12) and (4.13), (4.22) and (4.23) become:

$$I_{sxh} = \frac{1}{Z_{sh} + Z_{PPFh}} V_{sxh} + \frac{Z_{PPFh}}{Z_{sh} + Z_{PPFh}} I_{Lxh} \tag{4.24}$$

$$I_{cxh\_v} = \frac{1}{Z_{sh} + Z_{PPFh}} V_{sxh} + \frac{Z_{sh}}{Z_{sh} + Z_{PPFh}} I_{Lxh} \tag{4.25}$$

Compared with (4.22) and (4.23), as $K > 0$, it is obvious that when the LC-HAPF is operating, the harmonic current compensation capability is better than the case when only its pure PPF is used. Equation (4.22) also shows that the LC-HAPF can compensate the harmonic current caused by both the nonlinear load and the distorted system voltage. Moreover, it has the capability to suppress the parallel resonance between the PPF and the impedance of the power system, and also the series resonance of the PPF. In addition, the robustness of the system may also be improved when the LC-HAPF is adopted. In the following, the capabilities of the LC-HAPF to prevent both parallel and series resonance phenomena, to improve filtering effects of the PPF and to enhance the system robustness will be analyzed, discussed and verified by simulations.

4.4.2 Simulation Investigation of LC-HAPF Steady-State Compensating Performances

In the section, the LC-HAPF steady-state compensating performances are studied and discussed with: (1) capability to prevent the parallel resonance, (2) capability to prevent the series resonance, (3) capability to improve the filtering performance of the PPF and (4) capability to enhance the system robustness. Then the corresponding simulation results for the three-phase four-wire LC-HAPF as shown in Fig. 4.5 will be presented to illustrate and verify the PPF and LC-HAPF steady-state compensating performance analyses. The nonlinear load as shown in Fig. 4.5 is composed of three single-phase diode bridge rectifiers, which acts as harmonic producing load in the simulation. Simulation studies were carried out by using Power System Computer Aided Design (PSCAD)/Electro Magnetic Transient in DC System (EMTDC). Table 4.3 shows the simulated system parameters for analyzing the LC-HAPF steady-state compensating performances, where the coupling inductor L_{c1} and capacitor C_{c1} are tuned at 5th order harmonic frequency. To simplify the verification, the dc-link voltage of LC-HAPF is supported by external dc voltage source and the simulated three-phase loadings are balanced. In addition, the LC-HAPF reference compensating current can be determined by using the control block diagram as shown in Fig. 4.6. Then hysteresis PWM is applied for generating the required compensating current. Thus, the compensating current error Δi_{cx} together with hysteresis band H will be sent to the current PWM control part for generating the trigger signals to the switching devices (IGBTs) of the VSI.

4.4.2.1 LC-HAPF Capability to Prevent Parallel Resonance

The parallel resonance between the PPF and the power system impedance may occur when the operating conditions of the power system change, which results in a harmonic current amplification phenomenon. The magnitude diagrams of $K_{sxh\_i} = I_{sxh}/I_{Lxh}$ (4.16) with respect to frequency f and power system inductance Ls

Table 4.3 LC-HAPF simulated system parameters for its steady-state compensating performance analyses

System parameters	Physical values
$v_x(x = a, b, c), f$	220 V, 50 Hz
L_s, L_{c1}, C_{c1}	1, 5 mH, 80 μF
V_{dc1U}, V_{dc1L}	50 V
$L_{NL1x}, C_{NL1x}, R_{NL1x}$	30 mH, 200 μF, 26 Ω

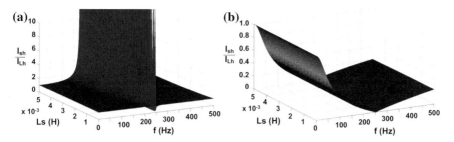

Fig. 4.10 Capability to prevent the parallel resonance between the PPF and the impedance of the power system: **a** when only PPF is utilized ($K = 0$) and **b** LC-HAPF is employed ($K = 50$)

when only PPF is employed or the LC-HAPF is employed are shown in Fig. 4.10, in which Fig. 4.10 can help to analyze the LC-HAPF capability to prevent parallel resonance. From Fig. 4.10a, when only pure PPF is used, $K = 0$ in (4.16), the harmonic current amplification phenomenon will occur from the f range of 177–250 Hz with Ls range from 0.1 to 5 mH respectively. When the LC-HAPF is employed, the harmonic current amplification phenomenon disappears, as shown in Fig. 4.10b. As a result, Fig. 4.10 clearly shows that the LC-HAPF has the capability to prevent the parallel resonance phenomenon, in which the pure PPF cannot avoid this phenomenon.

Figure 4.11 shows the corresponding simulation results for the parallel resonance prevention capability: (a) when only PPF is utilized and (b) LC-HAPF is employed. When Ls increases from 1 to 5 mH, the parallel resonance between the PPF and the power system impedance can occur close to the 3rd order harmonic frequency. This results in a 3rd order harmonic current amplification phenomenon at the system side. Figure 4.11a shows that the pure PPF compensation will enlarge the 3rd order harmonic current at the power system side from 2.01 to 4.59 A. This phenomenon deteriorates the compensation results as shown in Fig. 4.11a, in which the total harmonic distortion of the system current (THDisx = 54.5 %) and load voltage (THDvx = 10.5 %) do not satisfy the international standards (THDisx < 15 % for IEEE▲, THDisx < 20 % (I < 40 A) for Hong Kong Power Quality Standard, THDvx < 5 % for IEEE) [31–33]. Moreover, the system neutral current isn has been increased from 5.84 to 13.83 A with PPF compensation.

▲ With reference to the IEEE standard 519–2014 [31], the acceptable Total Demand Distortion (TDD) ≤ 15 % with I_{SC}/I_L is in $100 < 1000$ scale (a small rating 220 V – 10 kVA LC-HAPF prototype). The nominal rate current is assumed to be equal to the fundamental load current at the worst case analysis, which results in THD = TDD ≤ 15 %.

When the LC-HAPF is employed, Fig. 4.11b shows that the LC-HAPF can obtain better compensation results without the occurrence of the parallel resonance phenomenon. The amplitudes of system current fundamental, 3rd harmonic and other harmonic contents have been significantly reduced. The nonlinear system current and system displacement power factor (DPF) after the LC-HAPF

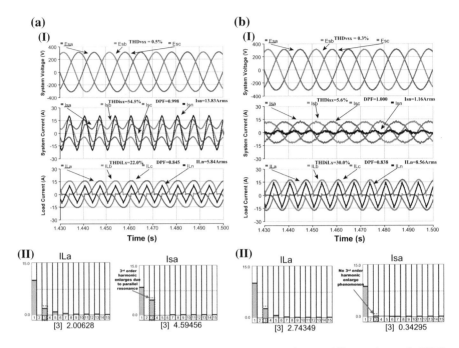

Fig. 4.11 Simulation results for the parallel resonance prevention capability: **a** when only PPF is utilized *i* Compensation performance, *ii* Frequency spectrum of phase a load current iLx and system current is x after compensation and **b** LC-HAPF is employed *i* Compensation performance, *ii* Frequency spectrum of phase a load current iLx and system current is x after compensation

Table 4.4 Simulation results before and after PPF and LC-HAPF compensation

Before compensation					After compensation			
Capabilities		THD_{isx} (%)	i_{sn} (A)	DPF	THD_{isx} (%)	i_{sn} (A)	DPF	THD_{vx} (%)
Prevent parallel resonance	PPF	22.0	5.84	0.845	54.5	13.83	0.998	10.5
	LC-HAPF	30.0	8.56	0.838	5.6	1.16	1.000	4.8
Prevent series resonance	PPF	30.0	8.10	0.838	>75.0	9.25	0.999	5.2
	LC-HAPF	30.0	8.56	0.838	8.5	1.17	1.000	4.6
Improve filtering performance	PPF	30.0	8.10	0.838	37.0	9.30	1.000	2.1
	LC-HAPF	30.0	8.56	0.838	5.0	0.90	1.000	2.2
Enhance system robustness	PPF	25.4	6.86	0.837	44.0	10.20	0.999	5.5
	LC-HAPF	30.0	8.56	0.838	5.2	1.05	1.000	4.0

compensation become sinusoidal (THD_{isx} = 5.6 %) and unity respectively, and the THD_{vx} of load voltage is 4.8 %, in which the LC-HAPF compensation results as summarized in Table 4.4 satisfy the international standards [31–33]. Moreover, i_{sn} has been reduced from 8.56 to 1.16 A after LC-HAPF compensation.

Figure 4.11 and Table 4.4 clearly illustrate that the LC-HAPF has the capability to prevent the parallel resonance, in which the pure PPF cannot avoid this phenomenon. These simulation results verify the previous PPF and LC-HAPF parallel resonance prevention analyses as shown in Fig. 4.10.

4.4.2.2 LC-HAPF Capability to Prevent Series Resonance

The PPF may fall into series resonance when the system harmonic voltage produces excessive harmonic current flowing into the PPF. The magnitude diagrams of $K_{cxh\_v} = I_{cxh}/V_{sxh}$ (4.21) with respect to f and L_s values when only PPF is employed or LC-HAPF is employed are shown in Fig. 4.12, which can help to analyze the LC-HAPF capability to prevent series resonance.

From Fig. 4.12a, when only pure PPF is used, $K = 0$ in (4.21), the PPF harmonic current amplification phenomenon due to the distorted system voltage V_{sxh} will occur from the f range of 177 to 250 Hz with Ls range from 0.1 to 5 mH respectively. When the LC-HAPF is employed, the PPF harmonic current amplification phenomenon disappears, as shown in Fig. 4.12b. As a result, Fig. 4.12 clearly illustrates that the LC-HAPF has capability to prevent the series resonance as well while the pure PPF cannot.

Figure 4.13 shows the corresponding simulation results for the series resonance prevention capability when: (a) only PPF is utilized and (b) LC-HAPF is employed. Since the coupling L_{c1} and C_{c1} are tuned at 5th order harmonic frequency as shown in Table 4.3, when 4 % of 5th order harmonic voltage is added to the system voltage, the series resonance of the PPF can occur close to the 5th order harmonic frequency. This results in a 5th order harmonic current amplification phenomenon at the system side. Figure 4.13a shows that the pure PPF compensation will enlarge the 5th order harmonic current at the power system side from 0.74 to 5.94 A. This phenomenon deteriorates the compensation results as shown in Table 4.4, in which the total harmonic distortion of the system current ($THD_{isx} > 75.0$ %) and load voltage ($THD_{vx} = 5.2$ %) do not satisfy the international standards [31–33]. Moreover, i_{sn} has been increased from 8.10 to 9.25 A with the PPF compensation.

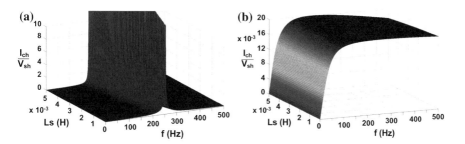

Fig. 4.12 Capability to prevent the series resonance between the PPF and the impedance of the power system: **a** when only PPF is utilized (K = 0) and **b** LC-HAPF is employed (K = 50)

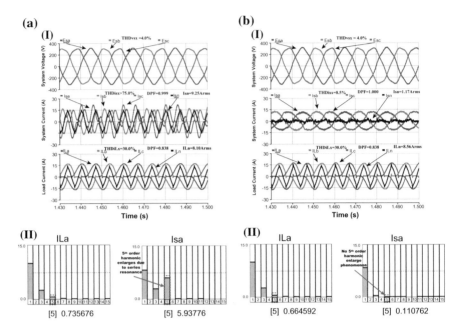

Fig. 4.13 Simulation results for the series resonance prevention capability: **a** when only PPF is utilized *i* Compensation performance, *ii* Frequency spectrum of phase a load current iLx and system current is x after compensation and **b** LC-HAPF is employed *i* Compensation performance, *ii* Frequency spectrum of phase a load current iLx and system current is x after compensation

When the LC-HAPF is employed, Fig. 4.13b shows that the LC-HAPF can obtain better compensation results without the occurrence of the series resonance phenomenon. The amplitudes of system current fundamental, 5th harmonic and other harmonic contents have been significantly reduced. The nonlinear system current and system DPF after LC-HAPF compensation become sinusoidal (THD_{isx} = 8.5 %) and unity respectively, and the THD_{vx} of load voltage is 4.6 %, in which the LC-HAPF compensation results as summarized in Table 4.4 satisfy the international standards [31–33]. Moreover, i_{sn} has been reduced from 8.56 to 1.17 A with the LC-HAPF compensation. Figure 4.13 and Table 4.4 clearly illustrate that the LC-HAPF has the capability to prevent parallel resonance while the pure PPF cannot. These simulation results verify the previous PPF and LC-HAPF series resonance prevention analyses as shown in Fig. 4.12.

4.4.2.3 LC-HAPF Capability to Improve Filtering Performance

From Table 4.3 and Fig. 4.14a shows a bode diagram of $K_{sxh\_i} = I_{sxh}/I_{Lxh}$ (4.16) with respect to different K values due to the effect of I_{Lxh}. Figure 4.14b shows a bode diagram of $K_{sxh\_v} = I_{sxh}/I_{Lxh}$ (4.20) with respect to different K values due to

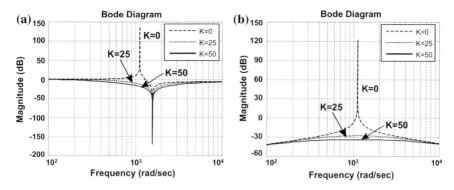

Fig. 4.14 Capability to improve the filtering performance of the PPF ($K = 0$, $K = 25$, $K = 50$) due to: **a** I_{Lxh} and **b** V_{sxh}

the effect of V_{sxh}. When only PPF is utilized, $K = 0$ in (4.16) and (4.20), the harmonic current amplification phenomenon occurs at $\omega = 1120$ rad/s. However, when the LC-HAPF is employed ($K = 25$ or $K = 50$), all the harmonic current components can be significantly decreased as shown in Fig. 4.14a, b. And the harmonic current amplification phenomenon also disappears. In addition, the larger the K value, the better the filtering performance can be obtained and vice versa. These results verify that the active inverter part of the LC-HAPF is capable of improving the filtering performance of the PPF and compensate the harmonic current caused by both I_{Lxh} and V_{sxh}.

Figure 4.15 shows the corresponding simulation results for the filtering capability when: (a) only PPF is utilized and (b) LC-HAPF is employed. Referring to Fig. 4.15a, when only pure PPF is employed, the amplitudes of the system current fundamental and 5th harmonic have been reduced. The system DPF becomes unity and the 5th harmonic current at the power system side reduces from 0.74 to 0.26 A. However, a larger THDisx is obtained after PPF compensation, from 30.0 to 37.0 %, which is due to the decrease of the system current fundamental component and increase of the 3rd order harmonic current at system side. The THD_{vx} of load voltage after PPF compensation is 2.1 %, in which the compensation results are summarized in Table 4.4. The THD_{isx} after PPF compensation does not satisfy the international standards [31, 33]. Moreover, i_{sn} has been increased from 8.10 to 9.30 A with PPF compensation.

When the LC-HAPF is employed, Fig. 4.15b shows that the amplitudes of the system current fundamental, 5th harmonic and other harmonic contents have been significantly reduced. The nonlinear system current and system DPF after LC-HAPF compensation become sinusoidal ($THD_{isx} = 5.0$ %) and unity respectively, and the THD_{vx} of load voltage is 2.2 %, in which the LC-HAPF compensation results as summarized in Table 4.4 satisfy the international standards [31–33]. Moreover, i_{sn} has been reduced from 8.56 to 0.90 A with LC-HAPF compensation. Figure 4.15 and Table 4.4 clearly illustrate that the LC-HAPF has

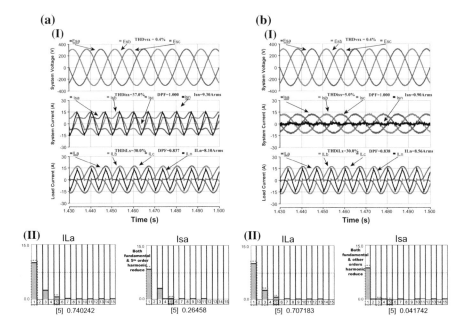

Fig. 4.15 Simulation results for improving the filtering effects of the PPF: **a** when only PPF is utilized *i* Compensation performance, *ii* Frequency spectrum of phase a load current iLx and system current is x after compensation and**b** LC-HAPF is employed *i* Compensation performance, *ii* Frequency spectrum of phase a load current iLx and system current is x after compensation

the capability of improving the filtering performance of the PPF. These simulation results verify the previous PPF and LC-HAPF filtering performance analyses as shown in Fig. 4.14.

4.4.2.4 LC-HAPF Capability to Enhance System Robustness

Due to the effect of I_{Lxh}, Fig. 4.16 shows a bode diagram of $K_{sxh\_i} = I_{sxh}/I_{Lxh}$ (4.16) with respect to different L_s values when only PPF or LC-HAPF is employed. When only the PPF is utilized, the harmonic current amplification point will move to the lower frequency side as the increase of L_s value and vice versa, as shown in Fig. 4.16a. Moreover, L_s value affects the system filtering performance. However, when the LC-HAPF is employed, the harmonic current amplification point disappears no matter what the L_s value is. Moreover, the harmonic current compensation characteristic of the LC-HAPF does not vary at all, as shown in Fig. 4.16b.

Due to the effect of V_{sxh}, Fig. 4.17 shows a bode diagram of $K_{sxh\_v} = I_{sxh}/V_{sxh}$ (4.20) with respect to different L_s values when only PPF or LC-HAPF is employed. Similar as Fig. 4.16, when only the PPF is utilized, harmonic current amplification point will move to the lower frequency side as the increase of L_s value and vice versa, as shown in Fig. 4.17a. However, when the LC-HAPF is employed, the

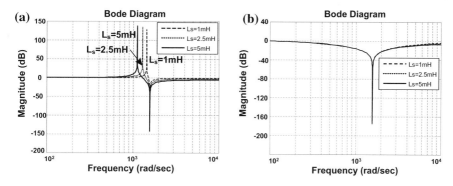

Fig. 4.16 Capability to enhance the system robustness due to I_{Lxh}: **a** when only PPF is utilized ($K = 0$) and **b** LC-HAPF is employed ($K = 50$)

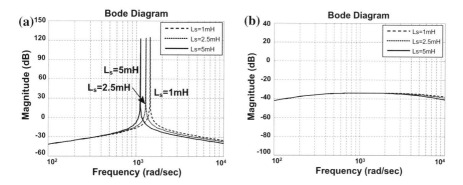

Fig. 4.17 Capability to enhance the system robustness due to V_{sxh}: **a** when only PPF is utilized ($K = 0$) and **b** LC-HAPF is employed ($K = 50$)

harmonic current amplification point disappears no matter what the L_s value is. Furthermore, the harmonic current compensation characteristic of the LC-HAPF does not vary at all, as shown in Fig. 4.17b. And it has also a large attenuation of more than -34 dB, which reflects I_{sxh} caused by V_{sxh} can be significantly reduced when the LC-HAPF is adopted. Both results in Figs. 4.16 and 4.17 verify that the LC-HAPF has the capability of enhancing the system robustness even though L_s is varying.

Figure 4.18 shows the corresponding simulation results for the system robustness enhancement capability when: (a) only PPF is utilized and (b) LC-HAPF is employed. When L_s increases from 1 to 3 mH, the harmonic current amplification point of the pure PPF will move to the lower frequency side. This results in enlarging the 3rd order harmonic current at the power system side as shown in Fig. 4.18a. Even though the amplitudes of the system current fundamental and 5th harmonic contents have been reduced, a larger THD_{isx} is obtained with PPF compensation, from 25.4 to 44.0 %. The increase of THD_{isx} is due to the decrease

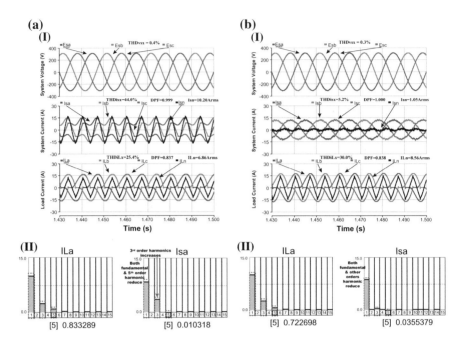

Fig. 4.18 Simulation results for enhancing the system robustness: **a** when only PPF is utilized *i* Compensation performance, *ii* Frequency spectrum of phase a load current iLx and system current is x after compensation and **b** LC-HAPF is employed *i* Compensation performance, *ii* Frequency spectrum of phase a load current iLx and system current is x after compensation

of the system current fundamental component and the increase of the 3rd order harmonic current. Compared with Figs. 4.15a and 4.18a yields a larger THD_{isx} with PPF compensation because increasing L_s to 3 mH moves the current amplification point closer to the 3rd order harmonic frequency, thus amplifying the amplitude of the 3rd order harmonic current more and deteriorating the compensating performance. The THD_{vx} of load voltage with PPF compensation is 5.5 %, in which the compensation results are summarized in Table 4.4. Neither THD_{isx} nor THD_{vx} with PPF compensation satisfy the international standards [31–33]. Moreover, i_{sn} has been increased from 6.86 to 10.20 A after PPF compensation.

When the LC-HAPF is employed, Fig. 4.18b shows that even though L_s increases from 1 to 3 mH, the LC-HAPF can still obtain good and similar compensation results compared with Fig. 4.15b. The amplitudes of system current fundamental, 5th harmonic and other harmonic contents have been significantly reduced. The nonlinear system current and system DPF after LC-HAPF compensation becomes sinusoidal (THD_{isx} = 5.2 %) and unity, and the THD_{vx} of load voltage is 4.0 %, in which the LC-HAPF compensation results as summarized in Table 4.4 satisfy the international standards [31–33]. Furthermore, i_{sn} has been reduced from 8.56 to 1.05 A with LC-HAPF compensation. Figure 4.18 and Table 4.4 clearly illustrate that the LC-HAPF has the capability of enhancing the system robustness. These simulation results verify the previous PPF and LC-HAPF in enhancing system robustness analyses as shown in Figs. 4.16 and 4.17.

The above simulation results and Table 4.4 clearly illustrate that the LC-HAPF is capable of preventing the parallel resonance between the PPF and the power system impedance, the series resonance of the PPF, improving the filtering characteristic of the PPF and enhancing the system robustness. Furthermore, the PPF and LC-HAPF steady-state compensating performance analyses are verified.

4.5 Dynamic Reactive Power Compensation and DC-Link Voltage Control Consideration for LC-HAPF

LC-HAPF is normally designed to deal with harmonic current rather than reactive power compensation [19–26], the inverter part is responsible to compensate harmonic currents only and the passive part provides a fixed reactive power value. In practical case, the load-side reactive power consumption usually varies from time to time, and if the loading mainly consists of induction motors such as centralized an air-conditioning system, its reactive power consumption will be much higher than the harmonic power consumption [34, 35]. As a result, it is necessary for the LC-HAPF [19–26] to perform dynamic reactive power compensation together with harmonic current compensation. Otherwise, the larger the reactive power compensation difference between the load-side and the coupling LC is, the larger the system current loss will be and it reduces the network efficiency. Moreover, there is not much detailed discussion on the LC-HAPF dc-link voltage control and also the inherent influence between the reactive power compensation and dc-link voltage control methods.

Due to the limitations among the existing literatures, this section aims to investigate and explore different dc-link voltage control techniques for the three-phase four-wire LC-HAPF while performing dynamic reactive power compensation:

(1) By using the traditional dc-link voltage control scheme as active current component, an extra start-up dc-link pre-charging control process is necessary [3, 30, 36, 37].

(2) To achieve the start-up dc-link voltage self-charging function, a dc-link voltage control as reactive current component for the LC-HAPF is proposed [38]; however, the LC-HAPF with this dc-link voltage control scheme fails to provide dynamic reactive power compensation.

(3) A novel dc-link voltage control method is proposed for achieving start-up dc-link self-charging function, dc-link voltage control and dynamic reactive power compensation. Moreover, the proposed method can be applied for the adaptive dc-link voltage reference control as will be discussed in Sect. 4.6.

Given that most of the loads in the distribution power systems are inductive, the following analysis and discussion will only focus on inductive loads [39].

4.5.1 Modeling of the DC-Link Voltage in a LC-HAPF Single-Phase Equivalent Circuit

From the three-phase four-wire center-spilt LC-HAPF as shown in Figs. 4.5 and 4.19 shows its single-phase equivalent circuit model. From Fig. 4.19, the compensating current i_{cx} can flow either C_{dc1U} or C_{dc1L} and returns through the neutral wire in both directions of IGBT switches. The dc-link capacitor voltages can be expressed as:

$$V_{dc1U} = \frac{1}{C_{dc1U}} \int i_{dc1U} dt, V_{dc1L} = -\frac{1}{C_{dc1L}} \int i_{dc1L} dt \qquad (4.26)$$

where i_{dc1U} and i_{dc1L} are the dc currents of upper and lower dc-link capacitor respectively, and

$$i_{dc1U} = s_x i_{cx}, i_{dc1L} = (1 - s_x) i_{cx} \qquad (4.27)$$

Substituting (4.27) into (4.26), the completed upper and lower dc capacitor voltages V_{dc1U} and V_{dc1L} become,

$$V_{dc1U} = \frac{1}{C_{dc1U}} \int s_x i_{cx} dt, V_{dc1L} = -\frac{1}{C_{dc1L}} \int (1 - s_x) i_{cx} dt \qquad (4.28)$$

$$s_x = \begin{cases} 1, & if \quad T_{1x} = 1 \quad T_{2x} = 0 \\ 0, & if \quad T_{1x} = 0 \quad T_{2x} = 1 \end{cases} \qquad (4.29)$$

In (4.28) and (4.29), S_x represents the switching function of one inverter leg in x phase based on the hysteresis current PWM method, which is the binary state of the upper and lower switches (T_{1x} and T_{2x}). When the positive direction of i_{cx} is assumed as shown in Fig. 4.19, the switching logic for each phase is formulated as

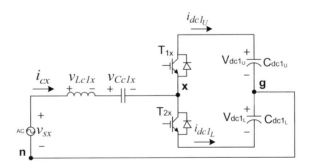

Fig. 4.19 Single-phase equivalent circuit model of a three-phase four-wire center-spilt LC-HAPF

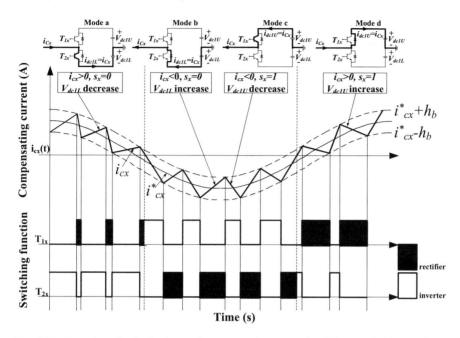

Fig. 4.20 Operation of a single-phase voltage source inverter under different switching modes by using hysteresis current PWM method

[40]: if $i_{cx} > (i^*_{cx} + h_b)$, T_{1x} is ON and T_{2x} is OFF, then $S_x = 1$; if $i_{cx} < (i^*_{cx} - h_b)$, T_{1x} is OFF and T_{2x} is ON, then $S_x = 0$, where i^*_{cx} is the reference compensating current, and h_b is the width of the hysteresis band.

According to the mathematical model in (4.28), if compensating current $i_{cx} > 0$, the upper dc-link voltage V_{dclU} is increased during switching function $S_x = 1$, while the lower dc-link voltage V_{dclL} is decreased for $S_x = 0$. The inverse results can be obtained if $i_{cx} < 0$. Figure 4.20 shows the operation of a single-phase voltage source inverter under different switching modes by using hysteresis current PWM method [41]. The changes of the upper and lower dc-link voltages V_{dclU} and V_{dclL} under different modes are summarized in Table 4.5.

Table 4.5 The change of the capacitor voltages (V_{dclU}, V_{dclL}) under different Modes

Switching mode	i_{cx} conditions	Switching function	Operating circuit	Change of dc capacitor voltage
a	$i_{cx} > 0$	$s_x = 0$, $T_{1x} = 0$, $T_{2x} = 1$	Inverter	V_{dclL} decrease
b	$i_{cx} < 0$	$s_x = 0$, $T_{1x} = 0$, $T_{2x} = 1$	Rectifier	V_{dclL} increase
c	$i_{cx} < 0$	$s_x = 1$, $T_{1x} = 1$, $T_{2x} = 0$	Inverter	V_{dclU} decrease
d	$i_{cx} > 0$	$s_x = 1$, $T_{1x} = 1$, $T_{2x} = 0$	Rectifier	V_{dclU} increase

4.5.2 Influence on DC-Link Voltage During LC-HAPF Performs Reactive Power Compensation

This section aims to present and analyze the influence on the dc-link voltage when LC-HAPF performs reactive power compensation. Through this analysis, the dc-link capacitor voltage will either be increased or decreased during fundamental reactive power compensation under insufficient dc-link voltage. Moreover, the influence is proportional to the difference between the LC-HAPF compensating current i_{cx} and its pure reactive reference i^*_{cxfq}, where the subscript 'f', 'p' and 'q' denote fundamental, active and reactive components.

4.5.2.1 Reactive Power Compensation Under Sufficient DC-Link Voltage

Under sufficient dc-link voltage, the compensating current generated by the LC-HAPF i_{cx} can track its pure reactive reference i^*_{cxfq} as shown in Fig. 4.21, thus the amplitude $I^*_{cxfq} \approx I_{cx}$. Provided that the hysteresis band is small enough, the PWM switching function in (4.29) will be evenly distributed as shown in Fig. 4.21. According to Table 4.5, the LC-HAPF will be changed between the operating modes of rectifier and inverter (modes a, b, c and d), and keeping the average dc-link voltage as a constant. In ideal lossless case, the dc-link voltage will not be affected when LC-HAPF performs reactive power compensation during $I^*_{cxfq} \approx I_{cx}$ case.

Fig. 4.21 i_{cx} and i^*_{cxfq} during $I^*_{cxfq} \approx I_{cx}$ case and the corresponding switching function

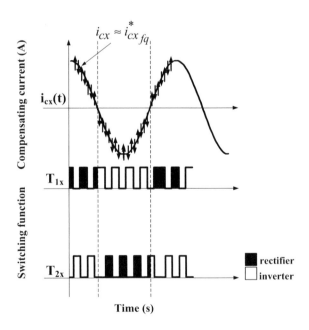

4.5.2.2 Reactive Power Compensation Under Insufficient DC-Link Voltage

Under insufficient dc-link voltage, the compensating reactive current generated by the LC-HAPF i_{cx} cannot track its pure reactive reference i^*_{cxfq}, in which there are two possible situations: (1) If the amplitude $I_{cx} > I^*_{cxfq}$, thus the instantaneous relationship gives $i_{cx} > i^*_{cxfq}$ during $i_{cx} > 0$, and $i_{cx} < i^*_{cxfq}$ during $i_{cx} < 0$ as shown in Fig. 4.22; (2) If the amplitude $I_{cx} < I^*_{cxfq}$, the opposite instantaneous relationship is given as shown in Fig. 4.23.

Hence, to force the compensating current i_{cx} to track its reference i^*_{cxfq} correspondingly:

(1) For $I_{cx} > I^*_{cxfq}$ case as shown in Fig. 4.22, when the hysteresis band is relatively small compared with the difference between i_{cx} and i^*_{cxfq}, their instantaneous relationship between i_{cx} and i^*_{cxfq} can be expressed as follow,

$$i_{cx} > (i^*_{cx_{fq}} + h_b) \quad \text{for} \quad i_{cx} > 0 \qquad (4.30)$$

$$i_{cx} < (i^*_{cx_{fq}} - h_b) \quad \text{for} \quad i_{cx} < 0 \qquad (4.31)$$

Fig. 4.22 i_{cx} and I^*_{cxfq} during $I_{cx} > I^*_{cxfq}$ case and the corresponding switching function

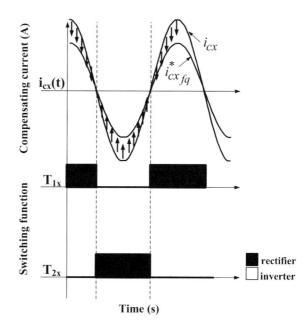

Fig. 4.23 i_{cx} and i^*_{cxfq} during $I_{cx} < I^*_{cxfq}$ case and the corresponding switching function

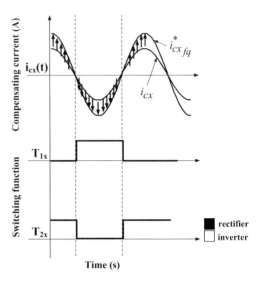

Based on the hysteresis PWM technique, the switching functions of (4.30) and (4.31) are,

$$s_x = 1(T_{1x} = 1, T_{2x} = 0) \quad \text{for} \quad i_{cx} > 0 \tag{4.32}$$

$$s_x = 0(T_{1x} = 0, T_{2x} = 1) \quad \text{for} \quad i_{cx} < 0 \tag{4.33}$$

According to Table 4.5, the PWM switching sequences in (4.32) and (4.33) drive the LC-HAPF to be operated in rectifier mode (modes b and d), thus increasing the average dc-link capacitor voltage. When the dc-link voltage is increased to sufficient high level, the LC-HAPF will change the operating mode from $I_{cx} > I^*_{cxfq}$ into $I^*_{cxfq} \approx I_{cx}$, and keeping the dc-link voltage value. Therefore, for a given fundamental reactive current reference i^*_{cxfq} ($I_{cx} > I^*_{cxfq}$), the dc-link voltage of LC-HAPF will be self-increased to a voltage level, which lets the reference compensating current can be tracked.

(2) For $I_{cx} < I^*_{cxfq}$ case as shown in Fig. 4.23, the instantaneous relationship between i_{cx} and i^*_{cxfq} is:

$$i_{cx} < (i^*_{cxfq} + h_b) \quad \text{for} \quad i_{cx} > 0 \tag{4.34}$$

$$i_{cx} > (i^*_{cxfq} - h_b) \quad \text{for} \quad i_{cx} < 0 \tag{4.35}$$

The switching functions of (4.34) and (4.35) are,

$$s_x = 0(T_{1x} = 0, T_{2x} = 1) \quad \text{for} \quad i_{cx} > 0 \tag{4.36}$$

$$s_x = 1(T_{1x} = 1, T_{2x} = 0) \quad \text{for} \quad i_{cx} < 0 \tag{4.37}$$

According to Table 4.5, the PWM switching sequences in (4.36) and (4.37) drive the LC-HAPF to be operated in inverter mode (modes a and c), thus decreasing the average dc-link capacitor voltage.

4.5.3   LC-HAPF Operation by Conventional DC-Link Voltage Control Methods

4.5.3.1   DC-Link Voltage Control Method as Active Current Component

Traditionally, if the indirect current (voltage reference) PWM control method is applied, the dc-link voltage of the inverter is controlled by feedback the dc control signal as reactive current component [3, 20–25, 30, 36, 37]. However, when the direct current PWM control method is applied in [3, 20–25, 30, 36, 37], the dc-link voltage should be controlled by feedback the dc control signal as active current component. Both dc-link voltage control methods are equivalent to each other.

When LC-HAPF performs reactive power compensating, the reference compensating current i_{cx}^* is composed by,

$$i_{cx}^* = i_{cxfq}^* + i_{cxfp\_dc}^* = i_{Lxfq}^* + i_{cxfp\_dc}^* \tag{4.38}$$

where $i_{cxfp\_dc}^*$ is the dc-link voltage controlled signal related to active current component, i_{Lxfq}^* is the loading fundamental reactive current which is equal to the reference compensating fundamental reactive current $i_{cxfq}^*, i_{cxfq}^* = i_{Lxfq}^*$.

However, to perform the reactive power and dc-link voltage control action in (4.38), a sufficient dc-link voltage should be provided to let the compensating current i_{cx} track with its reference i_{cx}^*. As a result, this conventional dc-link voltage control method fails to control the dc-link voltage during insufficient dc-link voltage, such as during start-up process. Due to this reason, when this dc-link voltage control method is applied in LC-HAPF, an extra start-up pre-charging control process is necessary. Usually, a three-phase uncontrollable rectifier is used to supply the initial dc-link voltage before operation [3, 30, 36, 37]. Moreover, when the adaptive dc-link voltage control idea that will be discussed in Sect. 4.6 is

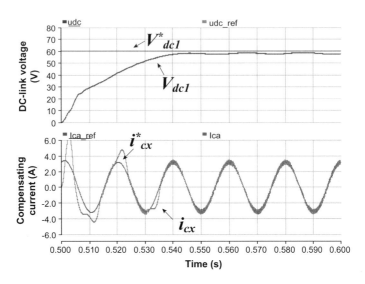

Fig. 4.24 V_{dcl}, i^*_{cx} and i_{cx} during $I_{cx} > I^*_{cxfq}$ condition with conventional dc-link voltage control method as active current component

applied, the reference dc-link voltage may be changed from a low level to a high level, at that occasion, the dc-link voltage may be insufficient to track the new reference value. Therefore, the conventional dc-link voltage control with pre-charging method may not work properly in the adaptive dc-link voltage controlled system.

The LC-HAPF in [20, 21, 23] showed that this dc-link voltage control can achieve start-up self-charging function without any external supply. That is actually due to the LC-HAPF in [20, 21, 23] is initially operating at $I_{cx} > i^*_{cxfq}$ condition. According to the analysis in Sect. 4.5.2, during $I_{cx} > i^*_{cxfq}$ condition, the dc-link voltage will be self-charging to a sufficient voltage level which lets the LC-HAPF's compensating current track with its reference $i^*_{cxfq} \approx i_{cx}$. Thus, the dc-link voltage can be maintained as its reference value in steady-state. However, if the LC-HAPF is initially operating at $I_{cx} < i^*_{cxfq}$ condition, this dc-link voltage control method fails to carry out this function.

Figures 4.24 and 4.25 show the simulation results of LC-HAPF start-up process by using the conventional dc-link voltage control method under different cases. From Fig. 4.24, when LC-HAPF starts operation during $I_{cx} > I^*_{cxfq}$ condition, the dc-link voltage V_{dcl} can carry out the start-up self-charging function and $i^*_{cx} \approx i_{cx}$ in steady-state. On the contrary, from Fig. 4.25, neither the dc-link voltage nor the reactive power compensation can be controlled when LC-HAPF is operating during $I_{cx} < I^*_{cxfq}$ condition, in which the simulation results verified the previous analysis.

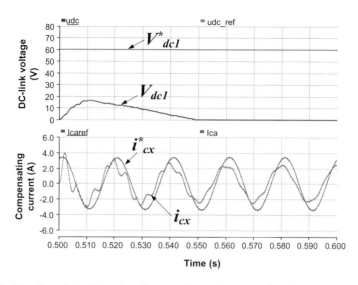

Fig. 4.25 V_{dcl}, i_{cx}^* and i_{cx} during $I_{cx} < I_{cxfq}^*$ condition with conventional dc-link voltage control method as active current component

4.5.3.2   DC-Link Voltage Control Method as Reactive Current Component

According to the analysis in Sect. 4.5.2, the dc-link voltage can be influenced by varying the reference compensating fundamental reactive current i_{cxfq}^* under insufficient dc-link voltage, thus the dc-link voltage can also be controlled by utilizing this influence. Hence, this dc-link voltage control method can achieve the start-up self-charging function. Moreover, it has been shown that this method is more effective than that conventional method as active current component [38]. However, when it is being applied in LC-HAPF, although the dc-link voltage can be controlled as its reference value, the LC-HAPF cannot perform dynamic reactive power compensation. The corresponding reason can be derived as follow:

By using the dc-link voltage control method as reactive current component, the reference compensating current i_{cx}^* is composed by,

$$i_{cx}^* = i_{cxfq}^* = i_{Lxfq}^* + i_{cxfq\_dc}^* \tag{4.39}$$

where $i_{cxfq\_dc}^*$ is the dc-link voltage control signal related to reactive current component.

Based on the analysis in Sect. 4.5.2, under insufficient dc-link voltage, the change of the dc-link voltage is direct proportional to the difference between the amplitude of I_{cx} and I_{cxfq}^*. Moreover, it has been concluded that for a given reference

i^*_{cxfq} ($I_{cx} < I^*_{cxfq}$), the dc-link voltage will be self-charging to a voltage level, then i_{cx} can trend its reference i^*_{cxfq} gradually, and maintaining this voltage level in steady-state. Therefore, to control the dc-link voltage V_{dc1} as its reference V^*_{dc1}, it will have a corresponding fixed reference value, that is $I_{cxfq} = I^*_{cxfq\_fixed}$. Therefore,

$$V^*_{dc1}\Big|_{I^*_{cxfq} = I^*_{cxfq\_fixed}} - V_{dc1} = 0 \tag{4.40}$$

Equation (4.40) implies that to control the dc-link voltage, the LC-HAPF must restrict to provide a fixed amount of reactive power. Therefore, by using the dc-link voltage control method as reactive current component, the LC-HAPF fails to perform dynamic reactive power compensation, in which the corresponding simulation and experimental results will be given in Sect. 4.5.5.

4.5.4 Proposed DC-Link Voltage Control Method

From the previous analysis, the dc-link voltage control method as active current component is effective only under sufficient dc-link voltage. In an insufficient dc-link voltage case, such as the LC-HAPF start-up process, both dc-link voltage control and reactive power compensation may fail. On the contrary, when the dc-link voltage control method as reactive current component is applied, the dc-link voltage control can be effectively controlled with start-up self-charging function. However, by using this control method, the LC-HAPF fails to perform dynamic reactive power compensation. As a result, a novel dc-link voltage control method by feedback the dc voltage controlled signal as both reactive and active current components is proposed in this section, so as to combine the advantages of both methods, which can achieve the start-up self-charging function, dc-link voltage control and dynamic reactive power compensation.

$$Q_{dc} = -k_q \cdot (V^*_{dc1} - V_{dc1}) \tag{4.41}$$

$$P_{dc} = k_p \cdot (V^*_{dc1} - V_{dc1}) \tag{4.42}$$

where Q_{dc} and P_{dc} are the dc control signals related to the reactive and active current components, k_q and k_p are the corresponding positive gains of the controller. By using the proposed control method, the reference compensating current i^*_{cx} is calculated by,

$$i^*_{cx} = i^*_{cxfq} + i^*_{cxfp\_dc} \tag{4.43}$$

where $i^*_{cxfq} = i^*_{Lxfq} + i^*_{cxfq\_dc}$.

In (4.42), P_{dc} represents the active power flow between the source and the LC-HAPF compensator, $P_{dc} > 0$ means the LC-HAPF absorb active power from the source, $P_{dc} < 0$ means the LC-HAPF inject active power to the source. According to the analysis in Sect. 4.5.2, the dc-link voltage will be charged for $I_{cx} > I^*_{cxfq}$, and discharged for $I_{cx} < I^*_{cxfq}$ during performing reactive power compensation. When $V^*_{dc1} - V_{dc1} > 0$, in order to increase the dc-link voltage, I^*_{cxfq} should be decreased by adding a negative Q_{dc}. On the contrary, when $V^*_{dc1} - V_{dc1} < 0$, in order to decrease the dc-link voltage, I^*_{cxfq} should be increased by adding a positive Q_{dc}. Therefore, the "$-$" sign is added in (4.41). In this book, in order to simplify the control process, Q_{dc} and P_{dc} in (4.41) and (4.42) are calculated by the same controller, i.e. $k_q = k_p$.

i^*_{Lxfq}, $i^*_{cxfq\_dc}$, $i^*_{cxfp\_dc}$ in (4.43) are calculated by using the three-phase instantaneous p-q theory [28]:

$$
\begin{bmatrix} i^*_{La_{fq}} \\ i^*_{Lb_{fq}} \\ i^*_{Lc_{fq}} \end{bmatrix} = \sqrt{\frac{2}{3}} \frac{1}{v_0 v^2_{\alpha\beta}} \begin{bmatrix} 1/\sqrt{2} & 1 & 0 \\ 1/\sqrt{2} & -1/2 & \sqrt{3}/2 \\ 1/\sqrt{2} & -1/2 & -\sqrt{3}/2 \end{bmatrix} \begin{bmatrix} v^2_{\alpha\beta} & 0 & 0 \\ 0 & v_0 v_\alpha & -v_0 v_\beta \\ 0 & v_0 v_\beta & v_0 v_\alpha \end{bmatrix} \begin{bmatrix} 0 \\ 0 \\ q_{\alpha\beta} \end{bmatrix}
$$

$$(4.44)$$

$$
\begin{bmatrix} i^*_{ca_{fq}\_dc} \\ i^*_{cb_{fq}\_dc} \\ i^*_{cc_{fq}\_dc} \end{bmatrix} = \sqrt{\frac{2}{3}} \frac{1}{v_0 v^2_{\alpha\beta}} \begin{bmatrix} 1/\sqrt{2} & 1 & 0 \\ 1/\sqrt{2} & -1/2 & \sqrt{3}/2 \\ 1/\sqrt{2} & -1/2 & -\sqrt{3}/2 \end{bmatrix} \begin{bmatrix} v^2_{\alpha\beta} & 0 & 0 \\ 0 & v_0 v_\alpha & -v_0 v_\beta \\ 0 & v_0 v_\beta & v_0 v_\alpha \end{bmatrix} \begin{bmatrix} 0 \\ 0 \\ Q_{dc} \end{bmatrix}
$$

$$(4.45)$$

$$
\begin{bmatrix} i^*_{ca_{fp}\_dc} \\ i^*_{cb_{fp}\_dc} \\ i^*_{cc_{fp}\_dc} \end{bmatrix} = \sqrt{\frac{2}{3}} \frac{1}{v_0 v^2_{\alpha\beta}} \begin{bmatrix} 1/\sqrt{2} & 1 & 0 \\ 1/\sqrt{2} & -1/2 & \sqrt{3}/2 \\ 1/\sqrt{2} & -1/2 & -\sqrt{3}/2 \end{bmatrix} \begin{bmatrix} v^2_{\alpha\beta} & 0 & 0 \\ 0 & v_0 v_\alpha & -v_0 v_\beta \\ 0 & v_0 v_\beta & v_0 v_\alpha \end{bmatrix} \begin{bmatrix} 0 \\ P_{dc} \\ 0 \end{bmatrix}
$$

$$(4.46)$$

where $q_{\alpha\beta}$ are the three-phase loading reactive power consumption, v_α, v_β and v_0 are the load voltages on the α-β-0 coordinate after the Clarke transformation [28], and $v_{\alpha\beta} = v^2_\alpha + v^2_\beta$.

Figure 4.26 shows the control process of the proposed dc-link voltage control method, when the LC-HAPF starts operation, the dc-link voltage is insufficient to let the compensating current i_{cx} track with its reference i^*_{cx}. Thus, the dc-link voltage control signal as reactive current component will dominate the control action in (4.43). During this period, the dc-link voltage will be self-charging under $I_{cx} > I^*_{cxfq}$ condition. As the dc-link voltage is increased and approaching the reference value, the compensating current tracking ability of the LC-HAPF will be improved gradually, and the control signals Q_{dc}, P_{dc} in (4.41) and (4.42) will also be decreased gradually, thus $i^*_{cx} \approx i_{cx}$ eventually. According to the analysis in

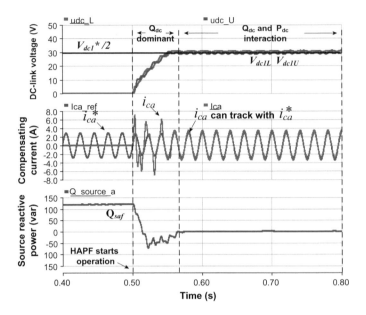

Fig. 4.26 Control process of the proposed dc-link voltage control method

Sect. 4.5.2, the dc-link voltage will not be affected by the reactive component when $i_{cx}^* \approx i_{cx}$. Hence, during this period, the dc-link voltage control signal as active current component will dominate the control action in (4.43). Therefore, the proposed dc-link voltage control method can be realized as:

(1) Q_{dc} control signal is used to step change the dc-link voltage under insufficient dc-link voltage, which can be effectively applied for start-up process and the adaptive dc-link voltage control that will be discussed in Sect. 4.6;

(2) P_{dc} control signal is used to maintain the dc-link voltage under sufficient dc-link voltage.

By using the proposed method, the LC-HAPF can compensate the reactive power consumed by the load, and keep the dc-link voltage as its reference one as shown in Fig. 4.26 ($Q_{sxf} \approx 0$var, $V_{dc1U} = V_{dc1L} = V_{dc1}^* = 30$ V). Therefore, the proposed dc-link voltage control method can effectively control the dc-link voltage without any extra pre-charging process and lets the LC-HAPF provide dynamic reactive power compensation. Table 4.6 shows the comparison among the conventional and proposed dc-link voltage control methods.

Figure 4.27 shows the control block diagram of the proposed dc-link voltage control method of the three-phase four-wire LC-HAPF, in which it consists of three main control blocks: (1) instantaneous power compensation control block, (2) proposed dc-link voltage control block, and (3) final reference compensating current PWM control block.

Table 4.6 Comparison between conventional and proposed dc-link control methods

Functions	DC-link voltage control methods		
	Feedback as active current component [3, 20–25, 30, 36, 37]	Feedback as reactive current component [38]	Proposed
Start-up self-charging	X	O	O
DC-link voltage control	O*	O	O
Adaptive dc-link voltage control	O*	O	O
Dynamic reactive power compensation	O*	X	O

Note O—function, X—failure, O*—conditionally can work under sufficient current tracking ability

(1) For the instantaneous power compensation control block, the loading fundamental reactive currents i^*_{Lxfq} are determined by the three-phase instantaneous p-q theory [28].

(2) The dc-link voltage is controlled by the proposed method (feedback dc controlled signal as both reactive and active current components). The dc-link voltage V_{dc1} is firstly obtained by summing up the measured upper and lower dc-link capacitor voltages (V_{dc1u} and V_{dc1L}). A low pass filter (LPF) is applied to filter out high frequency noise. Then the signal V_{dc1} is compared with the reference value V^*_{dc1}, and their difference will input to the proportional P controller to obtain the corresponding dc-link voltage control signals Q_{dc} and P_{dc}. If the proportional gains k_q, k_p in (4.41) and (4.42) are set too large, the stability of the control process will be degraded, and produces a large fluctuation during steady-state. On the contrary, if proportional gains are set too small, a long settling time and a large steady-state error will occur. Moreover, the dc-link voltage control method can also be applied by PI controller, the integral term can accelerate the movement of the process towards the reference value and eliminate the residual steady-state error. Since the parameter design of the dc-link voltage controller is not the main theme of this section, a pure proportional controller with an appropriate value is selected. A limiter is applied to avoid the overflow problem of the controller. After that, the final dc-link voltage control reference currents $i^*_{cxfq\_dc}, i^*_{cxfp\_dc}$ can be calculated, and they will be sent to current PWM control block to perform the dc-link voltage control.

(3) Then the final reference compensating current i^*_{cx} can be obtained by summing up the i^*_{Lxfq}, $i^*_{cxfq\_dc}$ and $i^*_{cxfp\_dc}$. Then i^*_{cx} together with compensating current i_{cx} will be sent to the current PWM control part for generating PWM trigger signals to control the power electronic switches of the inverter.

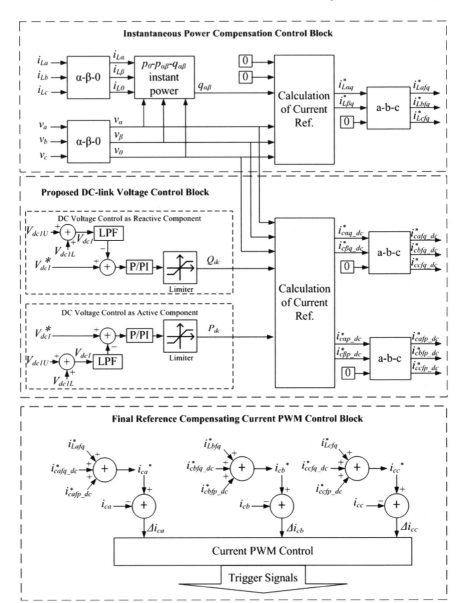

Fig. 4.27 Control block diagram for the three-phase four-wire LC-HAPF by using proposed dc-link voltage control method

4.5.5 Simulation and Experimental Verifications

To verify: (1) Failure dynamic reactive power compensation with dc-link voltage control as reactive current component; and (2) Effectiveness of the proposed dc-link control method. The simulated and experimental results in a small rating three-phase four-wire center-split LC-HAPF under balanced linear inductive loading will be given. Table 4.7 lists the simulated and experimental system parameters for the LC-HAPF as shown in Fig. 4.5. The coupling LC is designed based on: (1) an approximate mean value of the reactive power consumption for the 1st loading and both 1st and 2nd loadings, (2) the switching frequency with switching ripple less than 0.5 A with a maximum dc-link voltage of $V_{dc1}^*/2 = 40$ V. When the harmonic currents compensation is also taken consideration, the resonant frequency of the coupling LC should also be considered. Moreover, the dc-link voltage reference $V_{dc1}^*/2 = 30$ V in Table 4.7 is designed based on the minimum dc-link voltage requirement that will be discussed in Sect. 4.6. Simulation studies were carried out by using PSCAD/EMTDC. All control algorithms mentioned in this section are adopted in the LC-HAPF hardware prototype and implemented by a digital signal processor (DSP-TMSS320F2407).

Figure 4.28 shows the simulated and experimental reactive power at load side Q_{Lxf}. When the 1st inductive loading is connected, the simulated Q_{Lxf} for three-phase are 121.8 var with displacement power factor (DPF) = 0.786, while the three-phase experimental Q_{Lxf} are 116.0, 114.5 and 117.8 var with DPF = 0.804, 0.815 and 0.812 respectively. When both the 1st and 2nd inductive loadings are connected, the simulated Q_{Lxf} for three-phase increase to 176.6 var with DPF = 0.833, while the three-phase experimental Q_{Lxf} increase to 171.4, 168.6, 172.7 var with DPF = 0.835, 0.842 and 0.841 respectively.

4.5.5.1 Failure Dynamic Reactive Power Compensation with DC-Link Voltage Control as Reactive Current Component

With the implementation of the dc-link voltage control as reactive current component, Figs. 4.29 and 4.31 illustrate the whole simulated and experimental

Table 4.7 LC-HAPF system parameters for simulations and experiments

System parameters		Physical values
Source	V_x	55 V
	L_S	1 mH
LC-HAPF: (Reactive power supplied by passive part Q_{cxf} _PPF = −145.1 var)	L_c, C_c	6 mH, 140 μF
	$V_{dc1}^*/2$	30 V
1st inductive loading: $Q_{Lxf} = 121.8$ var	R_{L1}, L_{L1}	12 Ω, 30 mH
1st and 2nd inductive loadings: $Q_{Lxf} = 176.6$ var	R_{L1}, L_{L1}	12 Ω, 30 mH
	R_{L2}, L_{L2}	17.5 Ω, 30 mH

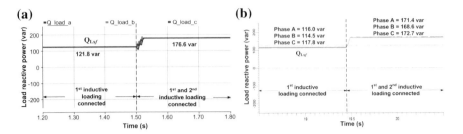

Fig. 4.28 Load-side fundamental reactive power: **a** simulated Q_{Lxf}, and **b** experimental Q_{Lxf}

dynamic reactive power compensation process for the loading situations as shown in Fig. 4.28, they include the waveforms of: (1) V_{dc1U}, V_{dc1L}, (2) Q_{cxf} of phase a, (3) Q_{sxf} of phase a. Figures 4.29a and 4.31a show that the simulation and experimental V_{dc1U}, V_{dc1L} level can be controlled as the reference value without any pre-charging process, and being kept at its reference 30 V no matter when the 1st inductive loading or 1st and 2nd loadings are connected, this verifies the effectiveness of the dc-link voltage control as reactive current component. Since the simulated and experimental loadings are approximately balanced, only phase a compensation diagrams will be illustrated. Even though the dc-link voltage can be controlled by this method, Figs. 4.29b and 4.31b illustrate that the simulated and experimental fundamental compensating reactive power Q_{cxf} are approximately fixed, no matter when the 1st or 1st and 2nd loadings are connected. That means the LC-HAPF fails to provide dynamic reactive power compensation by using this dc-link voltage control method. This result verifies the previous analysis in Sect. 4.5.3. As a result, the residual reactive power shown in Figs. 4.29c and 4.31c will be supplied by the source side with simulated and experimental Q_{sxf} of phase $a \approx -18.4, 45.1$ and $-23.5, 45.2$ var respectively when the 1st inductive loading or 1st and 2nd loadings are connected.

Tables 4.8 and 4.9 summarize the dynamic reactive power compensation results of the LC-HAPF based on the dc-link voltage control method with reactive current component. The three-phase simulated and experimental DPF of source-side can be improved when the 1st loading (or both the 1st and 2nd loadings) is connected. The simulated and experimental THD_{isx} are within 2.0 and 7.0 %. Figures 4.29, 4.31 and Tables 4.8 and 4.9 verify the previous analysis of the LC-HAPF failure in dynamic reactive power compensation by using conventional dc-link voltage control method as reactive current component.

4.5.5.2 Effectiveness of Proposed DC-Link Voltage Control Method

With the implementation of the proposed dc-link voltage control method, Figs. 4.30 and 4.32 illustrate the LC-HAPF whole simulated and experimental dynamic compensation process for the loading situations as shown in Fig. 4.28.

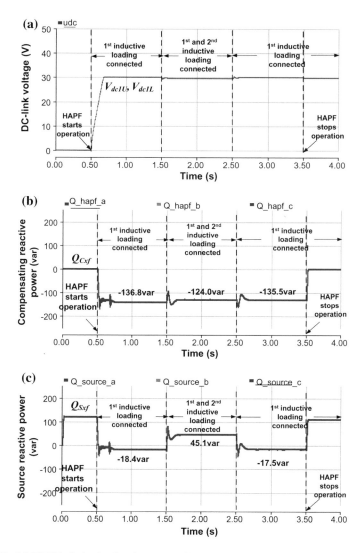

Fig. 4.29 *LC*-HAPF whole simulated process with dc-link voltage control method with reactive current component: **a** V_{dc1U}, V_{dc1L}, **b** Q_{cxf} of phase a, **c** Q_{sxf} of phase a

Figures 4.30a and 4.32a show that the simulation and experimental V_{dc1U}, V_{dc1L} level can also be kept at its reference 30 V with start-up self-charging function, this also verifies the effectiveness of the proposed dc-link voltage control method. Compared with Figs. 4.29b, 4.31b, 4.30b and 4.32b clearly illustrate that the LC-HAPF can inject different reactive power values, in which the simulated and experimental Q_{cxf} are varying with respect to different loading situations. As a result, the simulated and experimental Q_{sxf} can be approximately compensated close

Table 4.8 Simulation results before and after LC-HAPF reactive power compensation with dc-link voltage control method with reactive current component

Before compensation				After compensation			
Different cases	Q_{Lxf} (var)	DPF	i_{sx} (A)	Q_{sxf} (var)	DPF	i_{sx} (A)	THD_{isx} (%)
1st inductive loading	121.8	0.786	3.60	−18.4	0.991	2.90	1.2
1st and 2nd inductive loading	176.6	0.833	6.03	45.1	0.986	5.10	1.0

Table 4.9 Experimental results before and after LC-HAPF reactive power compensation with dc-link voltage control method with reactive current component

Before compensation					After compensation			
Different cases:		Q_{Lxf} (var)	DPF	i_{sx} (A)	Q_{sxf} (var)	DPF	i_{sx} (A)	THD_{isx} (%)
1st inductive loading	A	116.0	0.804	3.48	−23.5	0.974	2.98	6.9
	B	114.5	0.815	3.58	−30.3	0.972	3.07	6.5
	C	117.8	0.812	3.58	−22.6	0.974	2.90	5.9
1st and 2nd inductive loading	A	171.4	0.835	5.85	45.2	0.977	5.16	3.1
	B	168.6	0.842	5.91	42.1	0.980	5.22	3.5
	C	172.7	0.841	5.90	53.4	0.976	5.05	3.5

to zero no matter when the 1st loading or both 1st and 2nd loadings are connected, as shown in Figs. 4.30c and 4.32c, in which the simulated and experimental Q_{sxf} are significantly smaller than that of the dc-link voltage control method as shown in Figs. 4.29c and 4.31c.

Moreover, the proposed dc-link voltage control method can provide satisfactory dynamic response for LC-HAPF as shown in Figs. 4.33 and 4.34. In Fig. 4.33a and 4.34a, there is a period of time for which the source current waveforms are being settled, this time period is due to the LC-HAPF carrying out start-up self-charging process, the dc-link voltage is being charged from 0 V to its reference value $V_{dc1}^*/2 = 30$ V. Therefore, during the start-up process, the larger the dc-link voltage reference, the longer the source current settling time. After the dc-link voltage is controlled as the reference value, the LC-HAPF can have a fast dynamic response of less than one cycle after the 2nd loading is connected as shown in Fig. 4.33b and 4.34b.

Tables 4.10 and 4.11 summarize the dynamic reactive power compensation results of the LC-HAPF based on the proposed dc-link voltage control method. The three-phase simulated and experimental DPF of source-side can be further improved compared with the results of dc-link voltage control method as reactive current component. The simulated and experimental THD_{isx} are within 3.0 and 5.0 %. Figures 4.30, 4.32 and Tables 4.10 and 4.11 verify the effectiveness of the proposed dc-link voltage control method.

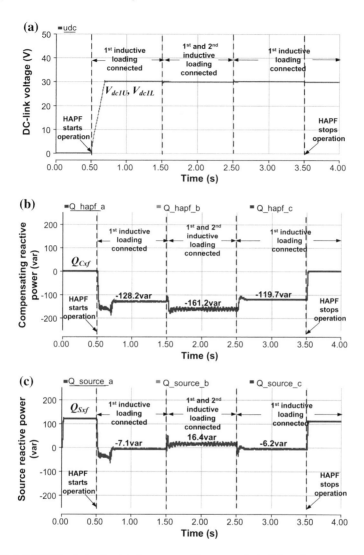

Fig. 4.30 LC-HAPF whole simulated process with proposed dc-link voltage control method: **a** V_{dc1U}, V_{dc1L}, **b** Q_{cxf} of phase a, **c** Q_{sxf} of phase a

In this section, as the LC-HAPF is tested under linear inductive loadings, there does not contain any harmonic components in source current i_{sx} in ideal case. The simulated and experimental THD_{isx} values are actually generated by the switching ripples with a fixed hysteresis band. Moreover, the large THD differences between the simulated and experimental results as shown in Tables 4.8, 4.9, 4.10 and 4.11 are actually due to the difference of component parameters, the resolution of the transducers, the signal conditional circuit error, the digital computation error and the noise in the experiment.

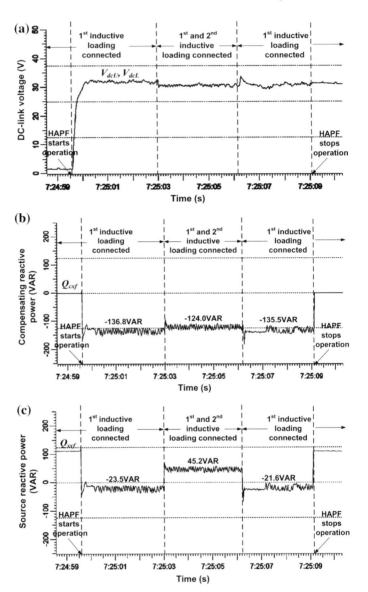

Fig. 4.31 LC-HAPF whole experimental process with dc-link voltage control method with reactive current component: **a** V_{dc1U}, V_{dc1L}, **b** Q_{cxf} of phase a, **c** Q_{sxf} of phase a

Compared the simulated and experimental results with the dc-link control method as reactive current component, the proposed method can: (1) also achieve the dc-link voltage control with start-up self-charging function; (2) provide dynamic reactive power compensation; (3) further improve the DPF; and (4) further reduce the RMS value of source current i_{sx}. As a result, it is clearly shown that LC-HAPF

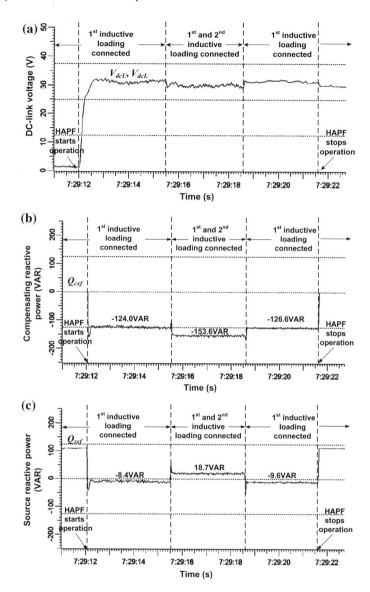

Fig. 4.32 LC-HAPF whole experimental process with proposed dc-link voltage control method: **a** V_{dc1U}, V_{dc1L}, **b** Q_{cxf} of phase a, **c** Q_{sxf} of phase a

adopting the proposed dc-link voltage control method can provide better compensating performances.

This section investigates different dc-link voltage control techniques for a three-phase four-wire center-split LC-HAPF during dynamic reactive power compensation. By using conventional dc-link voltage control method as active current

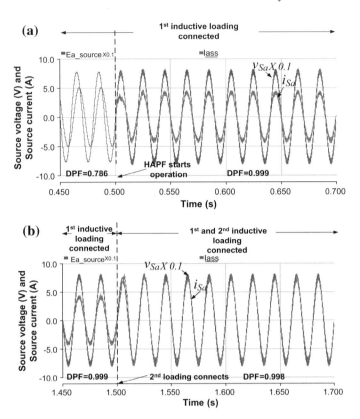

Fig. 4.33 Simulated dynamic response of LC-HAPF by using proposed dc-link voltage control method when: **a** the 1st loading is connected; **b** both the 1st and 2nd loadings are connected

component, an extra start-up dc-link pre-charging control process may be necessary. To achieve the start-up dc-link self-charging function, the dc-link voltage control as reactive current component can be applied; however, it fails to provide dynamic reactive power compensation. Through the proposed dc-link voltage control method:

(1) The LC-HAPF can achieve start-up dc-link self-charging function;
(2) The dc-link voltage of the LC-HAPF can be controlled as its reference level;
(3) The LC-HAPF can provide dynamic reactive power compensation;
(4) The adaptive dc-link voltage reference control that will be dscussed in Sect. 4.6 can be implemented.

Finally, simulation and experimental results of the three-phase four-wire center-spilt LC-HAPF under dynamic reactive power compensation application are presented to verify all discussions and analysis, and also show the effectiveness of the proposed dc-link voltage control method.

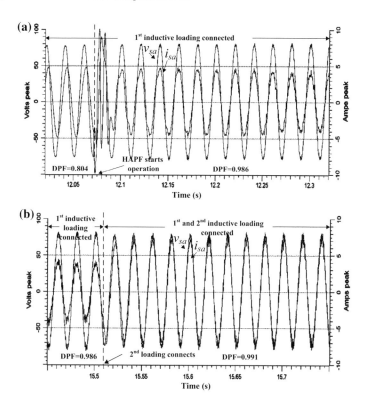

Fig. 4.34 Experimental dynamic response of LC-HAPF by using proposed dc-link voltage control method when: **a** the 1st loading is connected; **b** both the 1st and 2nd loadings are connected

Table 4.10 Simulation results before and after LC-HAPF reactive power compensation with proposed dc-link voltage control method

Before compensation				After compensation			
Different cases	Q_{Lxf} (var)	DPF	i_{sx} (A)	Q_{sxf} (var)	DPF	i_{sx} (A)	THD_{isx} (%)
1st inductive loading	121.8	0.786	3.60	−7.1	0.999	2.85	1.5
1st and 2nd inductive loading	176.6	0.833	6.03	16.4	0.998	5.01	2.4

Table 4.11 Experimental results before and after LC-HAPF reactive power compensation with proposed dc-link voltage control method

Before compensation					After compensation			
Different cases		Q_{Lxf} (var)	DPF	i_{sx} (A)	Q_{sxf} (var)	DPF	i_{sx} (A)	THD_{isx} (%)
1st inductive loading	A	116.0	0.804	3.48	−8.4	0.986	2.89	4.9
	B	114.5	0.815	3.58	−14.9	0.985	3.02	4.7
	C	117.8	0.812	3.58	−4.3	0.987	2.80	4.7
1st and 2nd inductive loading	A	171.4	0.835	5.85	18.7	0.991	5.05	2.9
	B	168.6	0.842	5.91	12.1	0.992	5.16	3.5
	C	172.7	0.841	5.90	24.5	0.989	4.93	3.5

4.6 Adaptive DC-Link Voltage Control Strategy for LC-HAPF and APF in Reactive Power Compensation

All LC-HAPF and other HAPFs discussed in [1–26] are based on fixed dc-link voltage reference. As the switching loss is directly proportional to the dc-link voltage [42], the system will obtain a larger switching loss if a higher dc-link voltage is used, and vice versa. Therefore, if the dc-link voltage can be adaptively changed according to different loading reactive power situations, the system can obtain better performance and operational flexibility.

To achieve the above functionality of the LC-HAPF, which is lacking among the existing literatures, an adaptive dc-link voltage control strategy for the three-phase four-wire LC-HAPF in dynamic reactive power compensation is given in this section. Compared with the traditional fixed dc-link voltage controlled LC-HAPF, the adaptive control strategy for the LC-HAPF can obtain less switching loss, switching noise and improve the compensating performance. Moreover, this control strategy can also be migrated into the APF application.

Given that most of the loads in the distribution power systems are inductive, the following analysis and discussion will only focus on the inductive loads [39]. Moreover, as this section mainly focuses on LC-HAPF reactive power compensation, only the reactive power compensation analysis, simulation and experimental results are included in this section.

4.6.1 Single-Phase Fundamental Equivalent Circuit Model of LC-HAPF

From the three-phase four-wire LC-HAPF circuit configuration as shown in Fig. 4.5, its single-phase fundamental equivalent circuit model is shown in Fig. 4.35, where the subscript 'f' denotes the fundamental frequency component. In the following analysis, all the parameters are in RMS values.

For simplicity, v_{sx} and v_x are assumed to be pure sinusoidal without harmonic components (i.e. $\vec{V}_x = \vec{V}_{xf} = |\vec{V}_x| = V_x$). From Fig. 4.35, the inverter fundamental voltage phasor $\vec{V}_{inv1xf}$ can be expressed as:

$$\vec{V}_{inv1xf} = \vec{V}_x - \vec{Z}_{PPFf} \cdot \vec{I}_{cxf}, \qquad (4.47)$$

where $\vec{Z}_{PPFf}$ is fundamental impedance of the PPF, the fundamental compensating current phasor $\vec{I}_{cxf}$ of the LC-HAPF can be expressed as $\vec{I}_{cxf} = I_{cxfp} + jI_{cxfq}$, the subscripts '$p$' and '$q$' denote the active and reactive components. I_{cxfp} is the fundamental active current for compensating loss and dc-link voltage control while

Fig. 4.35 LC-HAPF single-phase fundamental equivalent circuit model

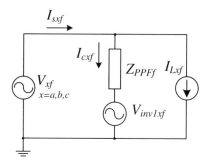

I_{cxfq} is the fundamental reactive current for the compensating reactive power of the loading. Simplify (4.47) yields,

$$\vec{V}_{inv1xf} = V_{inv1xfp} + jV_{inv1xfq} \tag{4.48}$$

where

$$\begin{aligned} V_{inv1xfp} &= \vec{V}_x + I_{cxfq}X_{PPFf} \\ V_{inv1xfq} &= -I_{cxfp}X_{PPFf} \end{aligned} \tag{4.49}$$

X_{PPFf} is the fundamental reactance of the PPF. From (4.49), the fundamental compensating active current I_{cxfp} and reactive current I_{cxfq} are:

$$I_{cxfp} = -\frac{V_{inv1xfq}}{X_{PPFf}} \tag{4.50}$$

$$I_{cxfq} = \frac{V_{inv1xfp} - V_x}{X_{PPFf}} \tag{4.51}$$

As the LC-HAPF aims to compensate fundamental reactive power, the steady-state active fundamental current I_{cxfp} from the inverter is small ($I_{cxfp} \approx 0$) when the dc-link voltage control is implemented. Thus, $V_{inv1xfq} \approx 0$. Therefore, the effect of dc-link voltage control for the LC-HAPF system can simply be neglected during steady-state situation.

For a fixed dc-link voltage level $V_{dc1U} = V_{dc1L} = 0.5V_{dc1}$ and modulation index m is assumed to be m ≈ 1, R_{Vdc} represents the ratio between the dc-link voltage V_{dc1U} and V_{dc1L} with the load voltage V_x reference to neutral n, it can be expressed as:

$$R_{Vdc} = \frac{\pm V_{inv1xf}}{V_x} = \frac{\pm 0.5V_{dc1}/\sqrt{2}}{V_x} = \pm\frac{V_{dc1}}{2\sqrt{2}V_x}, \tag{4.52}$$

where V_{inv1xf} is the inverter fundamental RMS voltage.

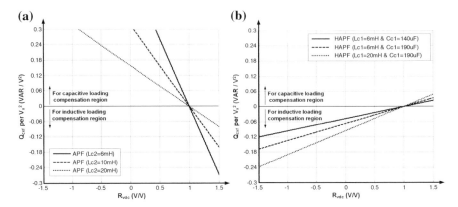

Fig. 4.36 Q_{cxf} per V_x^2 with respect to different R_{Vdc} for: **a** APF, and **b** LC-HAPF

The PPF is composed of a series connection of an inductor (L_{c1}) and a capacitor (C_{c1}), and C_{c1} dominates the passive part at fundamental frequency, it is a LC-HAPF system. However, if the PPF is changed to a pure inductor (L_{c2}), the original LC-HAPF system will become an APF. With the effect of dc-link voltage control being neglected ($I_{cxfp} = 0$) at steady-state, substituting $X_{PPFf} = X_{Lc2f}$ for APF and $X_{PPFf} = -|X_{Cc1f} - X_{Lc1f}|$ for LC-HAPF, their corresponding fundamental reactive power injection range Q_{cxf} per square of the load voltage level V_x^2 with respect to different R_{Vdc} can be shown in Fig. 4.36.

Since Q_{cxf} should be negative for inductive loading compensation, from Fig. 4.36, the ratio R_{Vdc} for APF must be at least greater than one, while the ratio R_{Vdc} for LC-HAPF can possibly be smaller than one while within a specific operational range. This means that the required V_{dc1U}, V_{dc1L} of APF must be larger than the peak of load voltage V_x regardless of the coupling inductance L_{c2}, while the V_{dc1U} and V_{dc1L} of LC-HAPF can be smaller than the peak of V_x within that operational range. When $R_{Vdc} = 0$, it means that both the APF and LC-HAPF are operating at pure passive filter mode, in which the APF at $R_{Vdc} = 0$ cannot support inductive loading compensation while the LC-HAPF can support a fixed Q_{cxf}. Furthermore, this fixed Q_{cxf} depends on the passive part parameters. Figure 4.36 clearly illustrates the main advantage of LC-HAPF over the traditional APF under inductive loading reactive power compensation. Under the same dc-link voltage consideration in Fig. 4.36b, when the coupling C_{c1} or L_{c1} of the LC-HAPF increases, the upper limit of $|Q_{cxf}|$ for inductive loading compensation region increases, on contrary, the lower limit of $|Q_{cxf}|$ for that region decreases and vice versa.

4.6.2 LC-HAPF Required Minimum DC-Link Voltage with Respect to Loading Reactive Power

Based on the previous assumption that the active fundamental current I_{cxfp} is very small ($I_{cxfp} \approx 0$) at steady-state, the inverter injects pure reactive fundamental current $\vec{I}_{cxf} = jI_{cxfq}$. Therefore, the $\vec{V}_{inv1xf}$ in (4.48) contains pure active part as $\vec{V}_{inv1xfp} = V_x - I_{cxfq}(X_{Cx1f} - X_{Lc1f})$ only. Then the LC-HAPF single-phase fundamental phasor diagram under inductive loading can be shown in Fig. 4.37. The vertical y-axis can be considered as the LC-HAPF active power (P/W) when locating $\vec{V}_x$ onto the LC-HAPF horizontal reactive power (Q/VAR) x-axis. The circle and its radius of $\vec{V}_{inv1xf\_dc} = \frac{0.5 \, V_{dc1}}{\sqrt{2}}$ represent the LC-HAPF fundamental compensation range and maximum compensation limit under a fixed dc-link voltage. $\vec{V}_{PPFxf}$ is the fundamental voltage phasor of the coupling LC. $\vec{I}_{Lxf}$ is the fundamental load current phasor, where I_{Lxfp} and I_{Lxfq} are the fundamental load active and reactive currents. In Fig. 4.37, the white semi-circle area represents LC-HAPF active power absorption region, whereas the shaded semi-circle area represents LC-HAPF active power injection region. When $\vec{V}_{inv1xf}$ is located inside the white semi-circle area, the LC-HAPF is absorbing active power, on the other hand, the LC-HAPF is injecting active power when $\vec{V}_{inv1xf}$ is located inside the shaded semi-circle area. When $\vec{V}_{inv1xf}$ is located onto the Q/VAR x-axis, the LC-HAPF does not absorb active power. From Fig. 4.37, the LC-HAPF reactive power compensation range with respect to different dc-link voltage can be deduced.

4.6.2.1 Full-Compensation by Coupling LC

When the loading reactive power Q_{Lxf} is fully compensated by coupling LC ($Q_{Lxf} = |Q_{cxf\_PPF}|$) as shown in Fig. 4.37a, the inverter does not need operation and output voltage ($V_{inv1xfp} = 0$). Thus, the switching loss and switching noise will be minimized in this situation. The LC-HAPF compensating reactive power Q_{cxf} is equal to the reactive power provided by the coupling LC Q_{cxf_PPF}:

$$Q_{cxf} = Q_{cxf\_PPF} = -\frac{V_x^2}{|X_{Cc1f} - X_{Lc1f}|} < 0 \tag{4.53}$$

where $Q_{cxf\_PPF} < 0$ means injecting reactive power or providing leading reactive power.

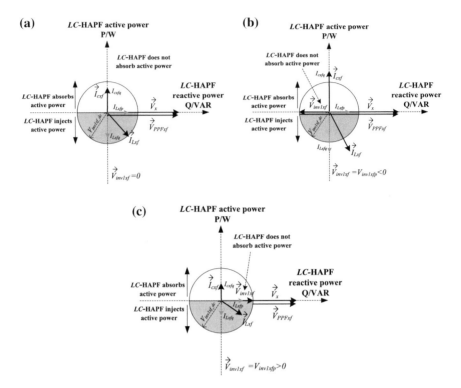

Fig. 4.37 LC-HAPF single-phase fundamental phasor diagram under inductive loading during: **a** full-compensation, **b** under-compensation and **c** over-compensation by coupling LC

4.6.2.2 Under-Compensation by Coupling LC

When the loading reactive power Q_{Lxf} is under-compensated by coupling LC $\left(Q_{Lxf} > |Q_{cxf\_PPF}|\right)$ as shown in Fig. 4.37b, in order to generate a larger I_{cxfq}, the inverter should output a negative inverter fundamental active voltage ($V_{inv1xfp} < 0$) as indicated by (4.51). With a fixed V_{dc1}, the LC-HAPF maximum compensating reactive power limit Q_{cxf_max} can be deduced through the under-compensation by coupling LC case, which can be expressed as:

$$Q_{cxf\_max} = -\frac{V_x^2(1 + |R_{Vdc}|)}{|X_{Cc1f} - X_{Lc1f}|} = Q_{cxf\_PPF}(1 + |R_{Vdc}|) < 0 \qquad (4.54)$$

4.6.2.3 Over-Compensation by Coupling LC

When the loading reactive power Q_{Lxf} is over-compensated by coupling LC $\left(Q_{Lxf} < |Q_{cxf\_PPF}|\right)$ as shown in Fig. 4.37c, in order to generate a smaller I_{cxfq}, the inverter should output a positive inverter fundamental active voltage ($V_{inv1xfp} > 0$) as indicated by (4.51). With a fixed V_{dc1}, the LC-HAPF minimum compensating reactive power limit Q_{cxf_min} can be deduced through the over-compensation by coupling LC case, which can be expressed as:

$$Q_{cxf\_max} = -\frac{V_x^2(1 - |R_{Vdc}|)}{|X_{Cc1f} - X_{Lc1f}|} = Q_{cxf\_PPF}(1 - |R_{Vdc}|) < 0 \quad (4.55)$$

From (4.54) and (4.55), the larger the dc-link voltage V_{dc1} or ratio R_{Vdc}, the larger the LC-HAPF compensation range can be obtained, and vice versa. However, a larger dc-link voltage will increase the LC-HAPF switching loss and generate a larger switching noise in the system, while a smaller dc-link will deteriorate the compensating performances if Q_{Lxf} is outside the LC-HAPF compensation range. When V_{dc1} is designed, the LC-HAPF reactive power compensating range for loading Q_{Lxf} can be expressed as:

$$|Q_{cxf\_min}| \leq Q_{Lxf} \leq |Q_{cxf\_max}| \quad (4.56)$$

Table 4.12 summarizes the LC-HAPF reactive power compensating range deduction steps under a fixed dc-link voltage $V_{dc1U} = V_{dc1L} = 0.5V_{dc1}$. When Q_{Lxf} is perfectly compensated by the coupling LC, the minimum dc-link voltage requirement ($V_{dc1U} = V_{dc1L} = 0$) can be achieved. In addition, the larger the reactive power compensation differences between the loading and the coupling LC, the larger the dc-link voltage requirement and vice versa. From Table 4.12, the required minimum dc-link voltage V_{dc1_minx} in each phase can be found by setting $Q_{Lxf} \approx |Q_{cxf\_min}| \approx |Q_{cxf\_max}|$ in (4.56),

$$V_{dc1\_minx} = 2\sqrt{2}V_x \left|1 - \frac{Q_{Lxf}}{|Q_{cxf\_PPF}|}\right| \quad (4.57)$$

Table 4.12 LC-HAPF reactive power compensation range deduction steps under a dc-link voltage $V_{dc1U} = V_{dc1L} = 0.5\ V_{dc1}$

1	With a fixed dc-link voltage $V_{dc1U} = V_{dc1L} = 0.5V_{dc1}$	
	$Q_{cxf\_PPF} = -\frac{V_x^2}{|X_{Cc1f} - X_{Lc1f}|} < 0$	(4.53)
	$Q_{cxf\_max} = Q_{cxf\_PPF}(1 + |R_{Vdc}|) < 0$	(4.54)
	$Q_{cxf\_min} = Q_{cxf\_PPF}(1 - |R_{Vdc}|) < 0$	(4.55)
	Where $R_{Vdc} = \left|\frac{V_{dc1}}{2\sqrt{2}V_x}\right|$	(4.52)
2	LC-HAPF reactive power compensating range for loading Q_{Lxf} $|Q_{cxf\_min}| \leq Q_{Lxf} \leq |Q_{cxf\_max}|$	(4.56)

in which (4.57) can be applied for the adaptive dc-link voltage control algorithm. Once the Q_{Lxf} is calculated, the corresponding V_{dc1_minx} in each phase can be obtained. Then the final three-phase required minimum dc-link voltage V_{dc1_min} can be chosen as follow:

$$V_{dc1\_min} = \max(V_{dc1\_mina}, V_{dc1\_minb}, V_{dc1\_minc}) \qquad (4.58)$$

In next section, the adaptive dc-link voltage controller for the three-phase four-wire LC-HAPF as shown in Fig. 4.5 will be given so that the LC-HAPF reactive power compensation range can be determined, switching loss and switching noise can be reduced, while comparing with the traditional fixed dc-link voltage LC-HAPF.

4.6.3 Adaptive DC-Link Voltage Controller for a Three-Phase Four-Wire LC-HAPF

Figure 4.38 shows the adaptive dc-link voltage control block diagram for the three-phase four-wire LC-HAPF in dynamic reactive power compensation, which consists of three main control blocks: the instantaneous power compensation control block, the adaptive dc-link voltage control block, and the final reference compensating current and PWM control block.

4.6.3.1 Instantaneous Power Compensation Control Block

For the instantaneous power compensation control block, the reference compensating currents for LC-HAPF (i_{cx_q}, the subscript $x = a$, b, c for three phases) are determined by the single-phase instantaneous p-q theory [29] as discussed in Sect. 4.3.2.

4.6.3.2 Adaptive DC-Link Voltage Control Block

The adaptive dc-link voltage control block consists of three parts: (1) the determination of adaptive minimum dc-link voltage V_{dc1_min}, (2) the determination of final reference dc-link voltage level V_{dc1}^* and (3) the dc-link voltage feedback P/PI controller.

(1) Determination of Adaptive Minimum DC-link Voltage: Initially, the loading instantaneous fundamental reactive power in each phase q_{Lxf} ($x = a$, b, c) can be calculated using single-phase instantaneous p-q theory [29] and low-pass filters. Usually, $-q_{Lxf}/2$ can keep as a constant value for more than one cycle, thus the loading fundamental reactive power consumption Q_{Lxf} in each phase

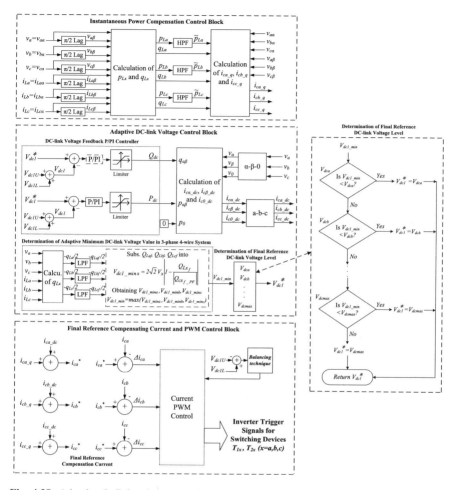

Fig. 4.38 Adaptive dc-link voltage control block diagram for the three-phase four-wire LC-HAPF

can be approximately treated as $Q_{Lxf} \approx -q_{Lxf}/2$. Then the required minimum dc-link voltage V_{dc1_minx} for compensating each phase Q_{Lxf} can then be calculated using (4.57), where V_x is the RMS load voltage and Q_{cxf_PPF}can be obtained according to (4.53). The adaptive minimum dc-link voltage will be equal to V_{dc1_min}, which can be determined by (4.58). To implement the adaptive dc-link voltage control function for the three-phase four-wire LC-HAPF, V_{dc1_min} can simply be treated as the final reference dc-link voltage V_{dc1}^*. It is obvious that as the loading reactive power consumption (Q_{Lxf}) changes, the system will adaptively yield different V_{dc1_minx} and V_{dc1_min} values.

(2) Determination of Final Reference DC-link Voltage Level: On contrary to the first part, this adaptive control scheme may frequently change the dc voltage

reference V_{dc1}^* in practical situations, because the loading is randomly determined by electric users (different Q_{Lxf}). Then this frequent change would cause
a rapid dc voltage fluctuation, resulting in deterioration of the LC-HAPF
operational performances [43]. In order to alleviate this problem, a final reference dc-link voltage level determination process is added as shown in
Fig. 4.38. The final reference dc-link voltage V_{dc1}^* is classified into certain
levels ($V_{dca}, V_{dcb}, \ldots, V_{dcmax}, V_{dca} < V_{dcb} \ldots < V_{dcmax}$) for selection, so that V_{dc1}^*
can be maintained as a constant value within a specific compensation range.
From Fig. 4.38, when the input V_{dc1_min} is less than the lowest dc voltage level
V_{dca}, the final reference dc-link voltage will be $V_{dc1}^* = V_{dca}$. Otherwise, the
steps should be repeated until V_{dc1_min} is found to be less than a dc-link
voltage level. However, if V_{dc1_min} is greater than the maximum voltage level
V_{dcmax}, the final reference dc-link voltage will be $V_{dc1}^* = V_{dcmax}$. In this way,
the dc-link voltage fluctuation problem under the adaptive dc voltage control
method can be lessened.

(3) DC-link Voltage Feedback P/PI Controller: The LC-HAPF can effectively
control the adaptive dc-link voltage and reactive power by providing feedback
of the dc-link voltage error signal as both reactive and active current components (Q_{dc}, P_{dc}) as discussed in previous Sect. 4.5. For the i_{cx} direction as
shown in Fig. 4.5, Q_{dc} and P_{dc} can be expressed as:

$$Q_{dc} = -k_q \cdot (V_{dc1}^* - V_{dc1}) \tag{4.59}$$

$$P_{dc} = k_p \cdot (V_{dc1}^* - V_{dc1}) \tag{4.60}$$

If the i_{cx} direction is opposite as the one shown in Fig. 4.5, the polarity of Q_{dc}
(4.59) and P_{dc} (4.60) will be inverse, in which Q_{dc} aims to change the dc-link
voltage level due to adaptive dc voltage control and start up the dc-link
self-charging function, while P_{dc} aims to maintain the dc-link voltage due to the
system loss. k_q and k_p are the corresponding positive proportional gains of the
controller. If the proportional gains k_q and k_p are set too large, the stability of the
control process will be degraded, and a large fluctuation during steady-state will be
produced. On the contrary, if they are set too small, a long settling time and a large
steady-state error will occur. To simplify the control process, Q_{dc} and P_{dc} in (4.59)
and (4.60) are calculated by the same controller, i.e. $k_q = k_p$, and an appropriate
value is selected. Actually, the adaptive control scheme can apply either P or PI
controller for the dc-link voltage control. Even though the P controller yields a
steady-state error, it is chosen because it is simpler and has less operational machine
cycles in the DSP, therefore it can yield a faster response when compared to the PI
controller. If the dc-link voltage with zero steady-state error is taken into consideration, PI controller is appreciated. A limiter is applied to avoid the overflow
problem of the controller. With the help of the three-phase instantaneous p-q theory

[27, 28], the dc-link voltage V_{dc1} can track its reference V_{dc1}^*, in which the details will be discussed in the following.

From [27, 28], the three-phase instantaneous load voltages (v_a, v_b, v_c) on the a-b-c coordinates can be transformed into those on the α-β-0 coordinates by the Clarke transformation:

$$\begin{bmatrix} v_0 \\ v_\alpha \\ v_\beta \end{bmatrix} = \sqrt{\frac{2}{3}} \begin{bmatrix} 1/\sqrt{2} & 1/\sqrt{2} & 1/\sqrt{2} \\ 1 & -1/\sqrt{2} & -1/\sqrt{2} \\ 0 & \sqrt{3}/2 & -\sqrt{3}/2 \end{bmatrix} \begin{bmatrix} v_a \\ v_b \\ v_c \end{bmatrix} \tag{4.61}$$

Then the three-phase dc voltage control reference compensating currents in α-β-0 coordinates can be calculated via (4.62):

$$\begin{bmatrix} i_{c0\_dc} \\ i_{c\alpha\_dc} \\ i_{c\beta\_dc} \end{bmatrix} = \frac{1}{v_0 v_{\alpha\beta}^2} \begin{bmatrix} v_{\alpha\beta}^2 & 0 & 0 \\ 0 & v_0 v_\alpha & -v_0 v_\beta \\ 0 & v_0 v_\beta & v_0 v_\alpha \end{bmatrix} \begin{bmatrix} 0 \\ P_{dc} \\ Q_{dc} \end{bmatrix} \tag{4.62}$$

where v_α, v_β, v_0 are the load voltages on the α-β-0 coordinate after the Clarke transformation, $v_{\alpha\beta}^2 = v_\alpha^2 + v_\beta^2$, P_{dc} and Q_{dc} are the feedback dc-link voltage controlled signals as active and reactive current components, which can be determined by (4.59) and (4.60). Finally, the three-phase dc voltage control reference compensating currents i_{cx_dc} in a-b-c coordinates can be obtained by the inverse matrix of Clarke transformation in α-β-0 coordinates, then the dc-link voltage V_{dc1} can track its reference V_{dc1}^* by changing the dc voltage reference compensating currents (i_{cx_dc}).

$$\begin{bmatrix} i_{ca\_dc} \\ i_{cb\_dc} \\ i_{cc\_dc} \end{bmatrix} = \sqrt{\frac{2}{3}} \begin{bmatrix} 1/\sqrt{2} & 1 & 0 \\ 1/\sqrt{2} & -1/2 & \sqrt{3}/2 \\ 1/\sqrt{2} & -1/2 & -\sqrt{3}/2 \end{bmatrix} \begin{bmatrix} i_{c0\_dc} \\ i_{c\alpha\_dc} \\ i_{c\beta\_dc} \end{bmatrix} \tag{4.63}$$

Therefore, the adaptive dc-link voltage control scheme for the LC-HAPF can then be implemented under various inductive linear loading conditions. In addition, the LC-HAPF initial start-up dc-link self-charging function can also be carried out by the adaptive dc-link voltage control scheme.

4.6.3.3 Final Reference Compensating Current and PWM Control Block

The hysteresis PWM or triangular carrier-based sinusoidal PWM method can be applied in the PWM control part. After the process of instantaneous power compensation and adaptive dc-link voltage control blocks as in Fig. 4.38, the final reference compensating current i_{cx}^* can be obtained by summing up the i_{cx_q} and i_{cx_dc}. Then the final reference and actual compensating currents i_{cx}^* and i_{cx} will be sent to the PWM control part, and the PWM trigger signals for the switching devices can then be generated. If the three-phase loadings are unbalanced, the dc capacitor voltage imbalance may occur, the dc capacitor voltage balancing concepts and techniques in [41] can be applied to balance the V_{dc1U} and V_{dc1L} under the adaptive dc voltage control method. The adaptive dc-link voltage controlled LC-HAPF can compensate the dynamic reactive power and reduce the switching loss and switching noise. In subsequent sections, the LC-HAPF simulated and experimental compensation results using the adaptive dc-link voltage control algorithm will be given, compared with a fixed dc controlled LC-HAPF.

4.6.4 Simulation and Experimental Verifications of the Adaptive DC-Link Voltage Controller for the Three-Phase Four-Wire LC-HAPF

Since this part aims to verify the adaptive dc-link voltage controlled LC-HAPF for reactive power compensation in a three-phase four-wire power system, the chosen 1st and 2nd testing loadings are linear inductive, so it is not necessary for the coupling LC to be tuned at a harmonic current order of the loading. Table 4.13 lists the simulated and experimental system parameters for the LC-HAPF. When the loading reactive power consumption is closed to that provided by the coupling LC of the LC-HAPF, the dc-link voltage requirement can be low. For $L_{c1} = 6$ mH, $C_{c1} = 140$ μF, the coupling LC supports a fixed reactive power of $|Q_{cxf\_PPF}| = 145.1$ var, in which the L_{c1} is designed based on the switching frequency with switching ripple less than 0.5 A under a maximum dc-link voltage of V_{dc1U}, $V_{dc1L} = 40$ V consideration.

The triangular carrier-based sinusoidal PWM method with current error limiter is applied so that the compensating current error must be within the triangular wave. And the frequency of the triangular wave is set to $f_{tri} = 7.5$ kHz. To simplify the verification, the simulated and experimental three-phase loadings are approximately balanced as shown in Fig. 4.39, so that the difference between V_{dc1U} and V_{dc1L} is

Table 4.13 LC-HAPF system parameters for simulations and experiments

System parameters		Physical values
V_x, f		55 V, 50 Hz
L_s		0.5 mH
$L_{c1} = 6$ mH, $C_{c1} = 140$ μF $|Q_{cxf\_PPF}| = 145.1$ var	Resistor R_{LL1x} and inductor L_{LL1x} for 1st inductive loading	10 Ω, 30 mH
	Resistor R_{LL2x} and inductor L_{LL2x} for 2nd inductive loading	22 Ω, 30 mH

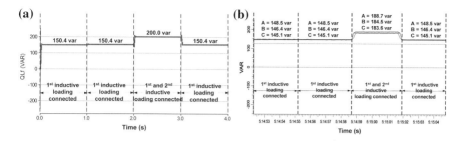

Fig. 4.39 Loading reactive power Q_{Lxf}: **a** simulated Q_{Lxf} and **b** experimental Q_{Lxf}

small ($V_{dc1U} \approx V_{dc1L}$) during the adaptive dc-link voltage control. As discussed before, to alleviate the dc-link voltage fluctuation problem under the adaptive control method, the final reference dc-link voltage V_{dc1}^* can be classified into certain levels for selection, in which V_{dc1}^* is set to have four levels (V_{dc1U}, V_{dc1L} = 10, 20, 30 and 40 V) for the following simulation and experimental verification.

Simulation studies were carried out by using PSCAD/EMTDC. In order to verify the simulation results, a 55 V/1.65 kVA three-phase four-wire LC-HAPF experimental prototype is applied. Figure 4.38 shows the adaptive dc-link voltage controlled LC-HAPF control block diagram for both simulations and experiments. Figure 4.39 shows the simulated and experimental reactive power at load-side Q_{Lxf}. When the 1st inductive loading is connected, the three-phase simulated Q_{Lxf} are 150.4 var with displacement power factor (DPF) = 0.729, while the three-phase experimental Q_{Lxf} are 148.5, 146.4 and 145.1 var with DPF = 0.763, 0.742 and 0.746 respectively. When both the 1st and 2nd inductive loadings are connected, the three-phase simulated Q_{Lxf} is increased to 200.0 var with DPF = 0.810, while the three-phase experimental Q_{Lxf} is increased to 188.7, 184.5, 183.6 var with DPF = 0.805, 0.815 and 0.822 respectively.

According to the designed four voltage levels (V_{dc1U}, V_{dc1L} = 10, 20, 30 and 40 V) for both simulations and experiments, from Table 4.14, the LC-HAPF required minimum dc-link voltage level will be V_{dc1U}, V_{dc1L} = 10 V or V_{dc1U},

Table 4.14 LC-HAPF simulated and experimental minimum dc-link voltage level with respect to Q_{Lxf} within V_{dc1U}, V_{dc1L} = 10, 20, 30 and 40 V

Reactive power by coupling LC Q_{cxf_PPF}	Simulated and experimental Q_{Lxf} for three phases (var)		Required $V_{dc1\_min}/2$ (V)	Minimum level V_{dc1U}, V_{dc1L} (V)
L_{c1} = 6 mH, C_{c1} = 140 μF Q_{cxf_PPF} = −145.1 var	1st loading	Sim.	150.4, 150.4, 150.4	2.9	10
		Exp.	148.5, 146.4, 145.1	1.8	10
	1st and 2nd loadings	Sim.	200.0, 200.0, 200.0	29.5	30
		Exp.	188.7, 184.5, 183.6	23.4	30

V_{dc1L} = 30 V for compensating the 1st loading or both the 1st and 2nd loadings. In the following section, the corresponding simulation and experimental results after the adaptive dc-link voltage controlled LC-HAPF compensation will be presented.

4.6.4.1 Adaptive DC-Link Voltage Controlled LC-HAPF

From Table 4.14, the required minimum dc-link voltage level for the LC-HAPF is V_{dc1U}, V_{dc1L} = 10 V for the 1st loading and V_{dc1U}, V_{dc1L} = 30 V for the 1st and 2nd loadings. Therefore, the adaptive control method for the LC-HAPF can reduce the switching loss and switching noise compared with a fixed V_{dc1U}, V_{dc1L} = 30 V case. For the simulated and experimental Q_{Lxf} as shown in Figs. 4.39, 4.40 and 4.41 illustrate the LC-HAPF simulated and experimental dynamic process of the adaptive dc-link voltage level and compensating performances. Figures 4.40a and 4.41a show that the simulated and experimental dc-link voltage level (V_{dc1U}, V_{dc1L}) can be adaptively changed according to different reactive power consumption of the loading. As the simulated and experimental loadings are approximately balanced, only phase b compensation diagrams will be illustrated. From Figs. 4.40b and 4.41b, the simulated and experimental system-side reactive power Q_{sxf} of phase b can be compensated close to zero regardless it is only the 1st loading or both 1st and 2nd loadings are connected. Compared with Figs. 4.39, 4.40c and 4.41c show that the simulated and experimental DPF of phase b can be compensated from 0.729 to 1.000 and 0.742 to 0.999 respectively once the LC-HAPF starts operation during the 1st inductive loading situation. From Figs. 4.40d and 4.41d, the simulated and experimental DPF of phase b can be kept at 0.990 or above when the 2nd loading is connected. The simulation results as shown in Fig. 4.40 are consistent with the experimental results as shown in Fig. 4.41.

Tables 4.15 and 4.16 summarize the simulation and experimental compensation results of the adaptive dc-link voltage controlled LC-HAPF. Compared with Q_{Lxf}, the three-phase simulated Q_{sxf} (5.0 or 22.0 var) and experimental Q_{sxf} (0.8, −1.5, 3.5 var or 25.2, 15.3, 22.2 var) have been compensated close to zero when the 1st loading or both 1st and 2nd loadings are connected. These can be verified by showing three-phase simulated and experimental DPF = 0.990 or above and THD_{isx} are within 3.0 and 5.0 % respectively. Moreover, the simulated and experimental i_{sx} can be significantly reduced after LC-HAPF compensation. Figures 4.40, 4.41 and Tables 4.15, 4.16 verify the effectiveness of the adaptive dc-link voltage controlled LC-HAPF for reactive power compensation.

4.6.4.2 Comparison Between Fixed and Adaptive DC-Link Voltage Controlled LC-HAPF

With a fixed dc-link voltage of V_{dc1U}, V_{dc1L} = 30 V, Figs. 4.42 and 4.43 show the LC-HAPF whole simulated and experimental dynamic compensation process for the same loading situations as shown in Fig. 4.39. Compared Figs. 4.42 and 4.43

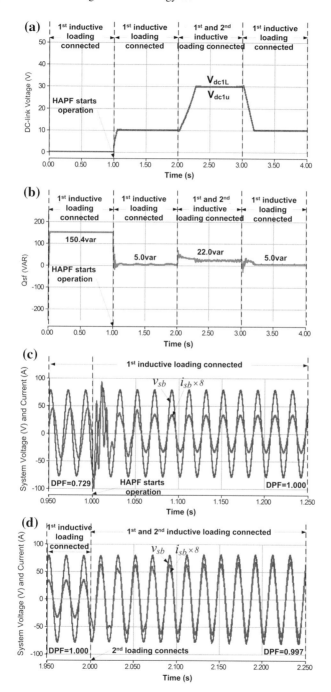

Fig. 4.40 LC-HAPF simulated dynamic process: **a** adaptive V_{dc1u}, V_{dc1L}, **b** Q_{sxf} of phase b, **c** DPF of phase b before and after LC-HAPF starts operation and **d** DPF of phase b before and after the 2nd loading is connected

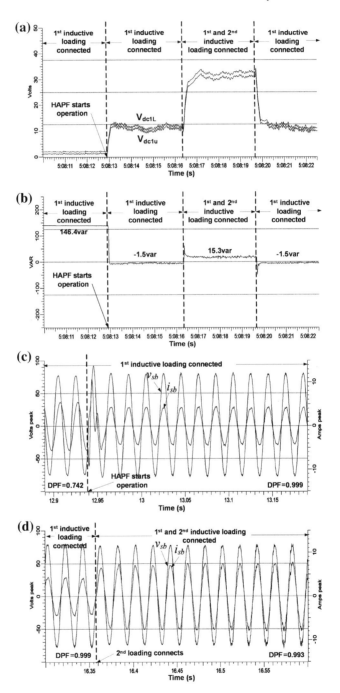

Fig. 4.41 LC-HAPF experimental dynamic process: **a** adaptive V_{dc1u}, V_{dc1L}, **b** Q_{sxf} of phase b, **c** DPF of phase b before and after LC-HAPF starts operation, and **d** DPF of phase b before and after the 2nd loading is connected

Table 4.15 Simulation results before and after LC-HAPF reactive power compensation with an adaptive dc-link voltage control

Before compensation				After compensation			
Different cases	Q_{Lxf} (var)	DPF	i_{sx} (A)	Q_{sxf} (var)	DPF	i_{sx} (A)	THD_{isx} (%)
1st inductive loading	150.4	0.729	4.01	5.0	1.000	2.95	2.4
1st and 2nd inductive loading	200.0	0.810	6.25	22.0	0.997	5.04	2.8

Table 4.16 Experimental results before and after LC-HAPF reactive power compensation with an adaptive dc-link voltage control

Before compensation					After compensation			
Different cases		Q_{Lxf} (var)	DPF	i_{sx} (A)	Q_{sxf} (var)	DPF	i_{sx} (A)	THD_{isx} (%)
1st inductive loading	A	148.5	0.763	4.06	0.8	0.999	3.04	3.4
	B	146.4	0.742	3.90	−1.5	0.999	3.05	4.9
	C	145.1	0.746	4.01	3.5	0.999	3.02	4.7
1st and 2nd inductive loading	A	188.7	0.805	6.03	25.2	0.990	4.97	3.3
	B	184.5	0.815	5.87	15.3	0.993	4.96	4.1
	C	183.6	0.822	5.94	22.2	0.991	4.90	4.7

with Figs. 4.40 and 4.41, the fixed and adaptive dc-link voltage control can achieve more or less the same steady-state reactive power compensation results. However, the adaptive control scheme just requires a lower dc-link voltage of V_{dc1U}, $V_{dc1L} = 10$ V for compensating the 1st loading as shown in Figs. 4.40a and 4.41a. For the adaptive dc-link voltage control of the LC-HAPF, due to the fact that its final reference dc-link voltage level V_{dc1}^{*} can vary according to different loading conditions, the compensating performance will be influenced during each changing of the dc voltage level. Compared with the fixed dc-link voltage control, the adaptive control scheme will have a longer settling time during the load and dc voltage level changing situation, which can be verified by the compensated Q_{sxf} as shown in Figs. 4.40 and 4.41 as well as Figs. 4.42 and 4.43.

Figures 4.44 and 4.45 show the LC-HAPF simulated and experimental compensating current i_{cx} of phase b and their spectra with (a) a fixed dc-link voltage of V_{dc1U}, $V_{dc1L} = 30$ V and (b) adaptive dc-link voltage control during the 1st loading connected case. As the reactive power consumption of the 1st inductive loading can almost be fully compensated by the coupling LC as shown in Table 4.14, it is clearly illustrated in Figs. 4.44 and 4.45 that the adaptive dc control scheme can effectively reduce the switching noise compared with the fixed V_{dc1U}, $V_{dc1L} = 30$ V.

Figures 4.39, 4.40, 4.41, 4.42, 4.43, 4.44, 4.45 and Tables 4.15, 4.16 verify that the adaptive dc-link voltage controlled LC-HAPF can obtain good dynamic reactive power compensating performance and reduce the system switching loss and switching noise, compared with the traditional fixed dc-link voltage LC-HAPF.

Fig. 4.42 LC-HAPF whole simulated dynamic process: **a** fixed V_{dc1u}, V_{dc1L} = 30 V, **b** Q_{sxf} of phase b, **c** DPF of phase b before and after LC-HAPF starts operation, and **d** DPF of phase b before and after the 2nd loading is connected

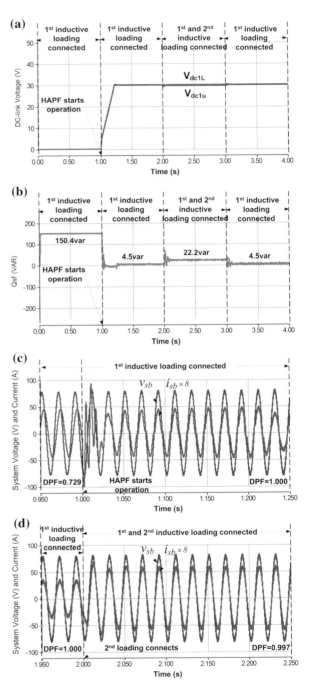

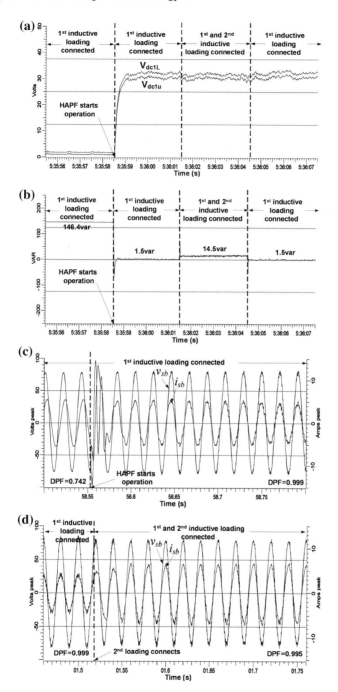

Fig. 4.43 LC-HAPF whole experimental dynamic process: **a** fixed V_{dc1u}, $V_{dc1L} = 30$ V, **b** Q_{sxf} of phase b, **c** DPF of phase b before and after LC-HAPF starts operation, and **d** DPF of phase b before and after the 2nd loading is connected

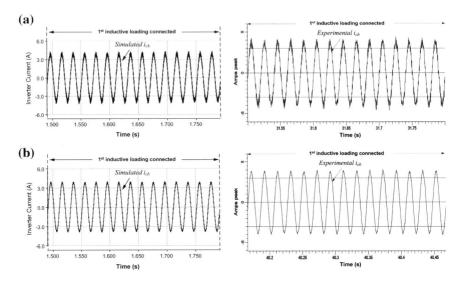

Fig. 4.44 Simulated and experimental i_{cx} of phase b with: **a** a fixed V_{dc1u}, $V_{dc1L} = 30$ V and **b** adaptive dc-link voltage control

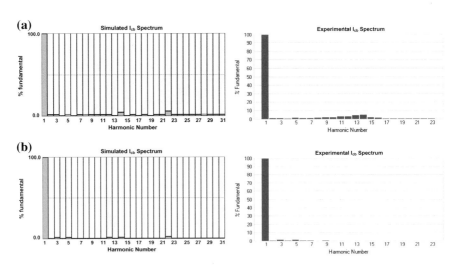

Fig. 4.45 Simulated and experimental frequency spectrum for i_{cx} of phase b with: **a** a fixed V_{dc1u}, $V_{dc1L} = 30$ V and **b** adaptive dc-link voltage control

4.6.5 Simulation and Experimental Verifications of the Adaptive DC-Link Voltage Controller for the Three-Phase Four-Wire APF

Actually, the adaptive dc-link voltage control idea can also be migrated into the APF system as well. From previous analysis, when the PPF is changed to a pure inductor (L_{c2}), the original LC-HAPF system will become an APF. By substituting the reactive power Q_{cxf_PPF} provided by the coupling L_{c2} into (4.57) and (4.58), the required minimum dc-link voltage V_{dc1_minx} in each phase and the final three-phase required minimum dc-link voltage V_{dc1_min} for the APF can be calculated. Then the adaptive dc-link voltage controller for the three-phase four-wire APF can be implemented.

Figure 4.46 illustrates the system configuration of a three-phase four-wire center-split APF. Table 4.17 lists the APF system parameters for simulations and experiments. As discussed before, to alleviate the dc-link voltage fluctuation problem under the adaptive control method, the final reference V_{dc1}^* is classified into three levels (V_{dc1U}, $V_{dc1L} = 200$, 250 and 300 V) for selection. To simplify the verification, the three-phase loadings are approximately balanced as shown in Figs. 4.47 and 4.48. Simulation studies were carried out by using PSCAD/EMTDC. In order to verify the simulation results, a three-phase four-wire center-split APF prototype is also designed and constructed. The control system of the prototype is a digital signal processor (DSP) TMS320F2812 and its sampling frequency is set at 25 kHz. Hysteresis current PWM is applied for the experimental prototype with maximum switching frequency, 12.5 kHz.

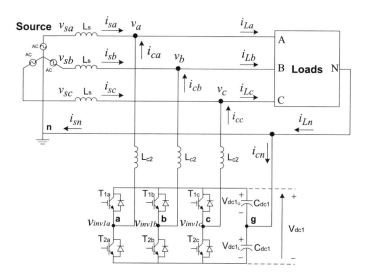

Fig. 4.46 Configuration of a three-phase four-wire center-split APF

Table 4.17 APF system parameters for simulations and experiments

System parameters	Physical values
V_x, f	110 V, 50 Hz
L_s, L_{c2}, C_{dc1}	0.5, 30 mH, 3.3 mF
Resistor R_{LL1x} and inductor L_{LL1x} for 1st inductive loading	17.0 Ω, 50.0 mH ($Q_{Lxf} \approx 355.0$ var)
Resistor R_{LL2x} and inductor L_{LL2x} for 2nd inductive loading	17.0 Ω, 50.0 mH ($Q_{Lxf} \approx 355.0$ var)

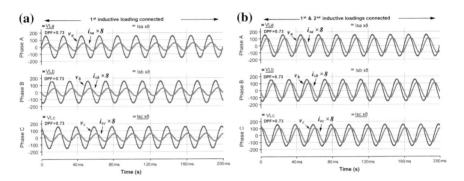

Fig. 4.47 Simulated displacement power factor (DPF) at source-side: **a** when 1st loading is connected and **b** when 1st and 2nd loadings are connected

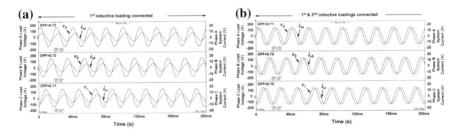

Fig. 4.48 Experimental displacement power factor (DPF) at source-side: **a** when 1st loading is connected and **b** when 1st and 2nd loadings are connected

Figures 4.47 and 4.48 show the simulated and experimental displacement power factor (DPF) at source-side under different loading cases. When the 1st loading is connected, the three-phase simulated DPF are 0.73, while the three-phase experimental DPF are 0.73, 0.72 and 0.71, respectively. When 1st and 2nd loading are connected, the three-phase simulated DPF are 0.73, while the three-phase experimental DPF are 0.71, 0.70 and 0.70, respectively.

Table 4.18 APF minimum dc-link voltage levels (200, 250 and 300 V)

Different situations		Required $V_{dc1\_min}/2$ (V)	Final minimum adapt. Level V_{dc1U}, V_{dc1L} (V)
1st loading	A, B, C	198.5	200
1st and 2nd loadings	A, B, C	241.5	250

According to the designed three voltage levels (V_{dc1U}, V_{dc1L} = 200 V, 250 V and 300 V) for the simulations and experiments, from Table 4.18, the APF required minimum dc-link voltage level will be V_{dc1U}, V_{dc1L} = 200 and 250 V for compensating the 1st loading, and 1st and 2nd loadings. In the following part, the corresponding simulation and experimental results after the proposed adaptive dc-link voltage controlled APF compensation will be presented in comparison with the conventional fixed dc-link voltage controlled APF.

4.6.5.1 Adaptive DC-Link Voltage Controlled APF

With the adaptive dc-link voltage control for the APF, Figs. 4.49 and 4.50 show the APF whole simulated and experimental dynamic compensation process for the loading situations as shown in Figs. 4.47 and 4.48. Figures 4.49 and 4.50 show that the simulated and experimental V_{dc1U}, V_{dc1L} can be adaptively changed according to different loading situations. Moreover, the simulated and experimental DPF can be improved to 0.99 or above after APF compensation for both loadings case, compared with Figs. 4.47 and 4.48.

Tables 4.19 and 4.20 summarize the simulated and experimental compensation results of the adaptive dc-link voltage controlled APF, the simulated and experimental system-side reactive power Q_{sxf} can be compensated close to zero when the 1st loading, and both the 1st and 2nd loadings are connected. These results can be verified by showing the three-phase simulated and experimental DPF = 0.99 or above and THD_{isx} are within 5.0 % respectively. Moreover, the system current i_{sx} can be significantly reduced after the APF compensation. Figures 4.49 and 4.50 and Tables 4.19 and 4.20 verify the effectiveness of the proposed adaptive dc-link voltage controlled APF for reactive power compensation.

4.6.5.2 Comparison Between Fixed and Adaptive DC-Link Voltage Controlled APF

With a fixed dc-link voltage of V_{dcU}, V_{dcL} = 300 V, Figs. 4.51 and 4.52 show the APF whole simulated and experimental dynamic compensation process for the same loading situations as shown in Figs. 4.47 and 4.48. Tables 4.21 and 4.22

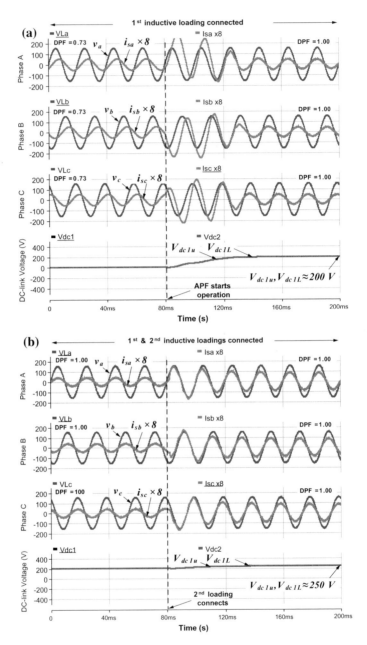

Fig. 4.49 APF simulated dynamic process with an adaptive dc-link voltage control: **a** DPF and V_{dc1U}, V_{dc1L} before and after APF starts operation during 1st loading and **b** DPF and V_{dc1U}, V_{dc1L} when 2nd loading is connected

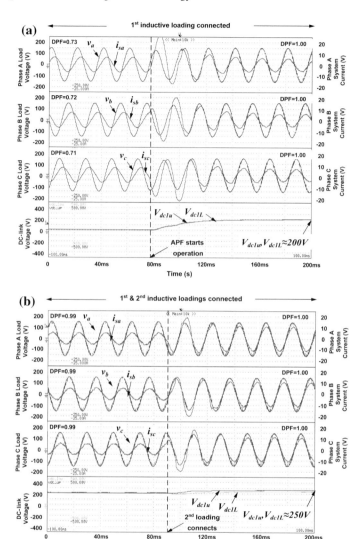

Fig. 4.50 APF experimental dynamic process with an adaptive dc-link voltage control: **a** DPF and V_{dc1U}, V_{dc1L} before and after APF starts operation during 1st loading and **b** DPF and V_{dc1U}, V_{dc1L} when 2nd loading is connected

summarize the simulated and experimental compensation results of the fixed dc-link voltage (V_{dcU}, V_{dcL} = 300 V) controlled APF correspondingly.

From Figs. 4.49, 4.50, 4.51, 4.52 and Tables 4.19, 4.20, 4.21, 4.22, the fixed and adaptive dc-link voltage controlled APF can achieve more or less the same steady-state reactive power compensation results. However, the adaptive control scheme just requires a lower dc-link voltage for compensating the 1st loading and

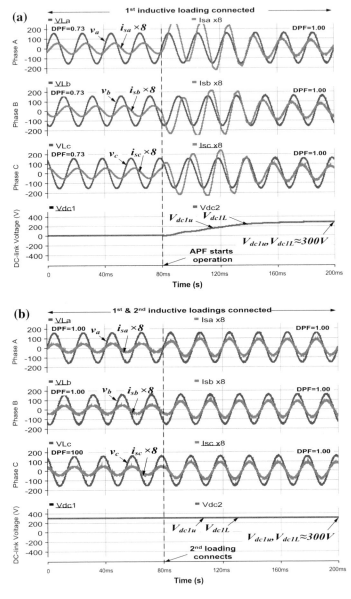

Fig. 4.51 APF simulated dynamic process with a fixed V_{dc1U}, $V_{dc1L} = 300$ V: **a** DPF and V_{dc1U}, V_{dc1L} before and after APF starts operation during 1st loading and **b** DPF and V_{dc1U}, V_{dc1L} when 2nd loading is connected

1st and 2nd loadings as shown in Figs. 4.49, 4.50, 4.51 and 4.52. Thus, it obtains less THD_{isx} and i_{sx} values for both loading cases, because a lower dc-link voltage will generate less switching noise into the system, and vice versa. For the adaptive

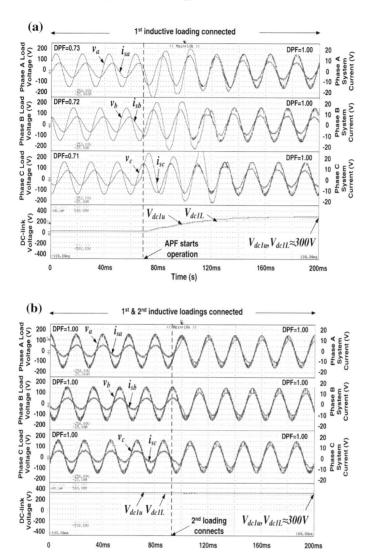

Fig. 4.52 APF experimental dynamic process with a fixed V_{dc1U}, $V_{dc1L} = 300$ V: **a** DPF and V_{dc1U}, V_{dc1L} before and after APF starts operation during 1st loading and **b** DPF and V_{dc1U}, V_{dc1L} when 2nd loading is connected

dc-link voltage controlled APF, due to its final reference V_{dc1}^* can be varied according to different loading conditions, the compensating performance will be influenced during each changing of the dc voltage level. Compared with the fixed dc-link voltage control, the proposed one yields a longer settling time during the load and dc voltage level changing situation.

Table 4.19 Simulation results before and after APF reactive power compensation with an adaptive dc-link voltage control

Before compensation				After compensation			
Different cases	Q_{Lxf} (var)	DPF	i_{sx} (A)	Q_{sxf} (var)	DPF	i_{sx} (A)	THD_{isx} (%)
1st inductive loading	350	0.73	4.73	18	1.00	3.61	4.9
1st and 2nd inductive loading	700	0.73	9.46	11	1.00	7.15	3.0

Table 4.20 Experimental results before and after APF reactive power compensation with an adaptive dc-link voltage control

Before compensation					After compensation			
Different cases		Q_{Lxf} (var)	DPF	i_{sx} (A)	Q_{sxf} (var)	DPF	i_{sx} (A)	THD_{isx} (%)
1st inductive loading	A	354	0.73	4.76	69	0.99	3.65	5.0
	B	363	0.72	4.88	69	0.99	3.65	4.8
	C	357	0.71	4.81	66	0.99	3.58	4.9
1st and 2nd inductive loading	A	702	0.71	9.46	78	1.00	7.39	4.4
	B	686	0.70	9.47	76	1.00	7.24	4.5
	C	670	0.70	9.44	75	1.00	7.25	4.1

Table 4.21 Simulation results before and after APF reactive power compensation with a fixed V_{dc1U}, V_{dc1L} = 300 V

Before compensation				After compensation			
Different cases	Q_{Lxf} (var)	DPF	i_{sx} (A)	Q_{sxf} (var)	DPF	i_{sx} (A)	THD_{isx} (%)
1st inductive loading	350	0.73	4.73	−8	1.00	3.87	7.5
1st and 2nd inductive loading	700	0.73	9.46	5	1.00	7.28	4.0

Table 4.22 Experimental results before and after APF reactive power compensation with a fixed V_{dc1U}, V_{dc1L} = 300 V

Before compensation					After compensation			
Different cases		Q_{Lxf} (var)	DPF	i_{sx} (A)	Q_{sxf} (var)	DPF	i_{sx} (A)	THD_{isx} (%)
1st inductive loading	A	354	0.73	4.76	42	1.00	3.81	6.3
	B	363	0.72	4.88	41	1.00	3.76	6.3
	C	357	0.71	4.81	40	1.00	3.70	6.3
1st and 2nd inductive loading	A	702	0.71	9.46	81	1.00	7.58	4.6
	B	686	0.70	9.47	79	1.00	7.42	5.0
	C	670	0.70	9.44	77	1.00	7.40	4.5

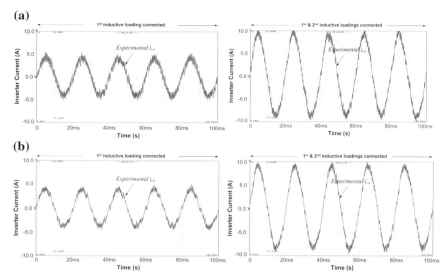

Fig. 4.53 Experimental i_{cx} of phase a with: **a** a fixed V_{dcU}, V_{dcL} = 300 V and **b** proposed adaptive dc-link voltage control

Figure 4.53 shows the APF experimental compensating current i_{cx} of phase a with: (a) fixed V_{dc1U}, V_{dc1L} = 300 V and (b) proposed adaptive dc-link voltage control during 1st loading and 1st and 2nd loadings cases. Figure 4.53 shows that the adaptive dc-link voltage control scheme can effectively reduce the switching noise of the APF compared with the traditional fixed V_{dc1U}, V_{dc1L} = 300 V operation. Moreover, from Table 4.23, the proposed adaptive control scheme can reduce the inverter power loss by 47 and 19 %, respectively compared with the conventional fixed V_{dc1U}, V_{dc1L} = 300 V control.

From Figs. 4.47, 4.48, 4.49, 4.50, 4.51, 4.52, 4.53 and Tables 4.19, 4.20, 4.21, 4.22, 4.23, they verified that the proposed adaptive dc-link voltage controlled APF can obtain good dynamic reactive power compensating performance and reduce the system switching loss and switching noise.

Soft-switching techniques are usually applied to reduce the system operating switching loss. However, they all require extra auxiliary circuits, thus increasing the system initial cost. To obtain loss reduction function without adding extra circuit

Table 4.23 Experimental inverter power loss of APF with a fixed V_{dc1U}, V_{dc1L} = 300 V and adaptive dc-link voltage control

Inverter power loss of APF		Fixed V_{dcU}, V_{dcL} = 300 V (W)	Adaptive dc
Power loss (W)	1st loading	163.2	86.4 W (200 V) ~47 %↓
	1st and 2nd loading	213.0	172.2 W (250 V) ~19 %↓

components and to implement the dynamic reactive power compensation capability for the LC-HAPF, an adaptive dc-link voltage controlled LC-HAPF for dynamic reactive power compensation capability is given in this section. Moreover, its viability and effectiveness have been proved by simulation and experimental results, in which it can achieve a good dynamic reactive power compensating performance as well as reducing the switching loss and switching noise compared with the traditional fixed dc-link voltage controlled LC-HAPF. After that, the adaptive dc-link voltage control idea is also extended into the APF application, in which the simulation and experimental results also show significantly reduction in both switching loss and switching noise compared with the traditional fixed dc-link voltage controlled APF.

4.7 Minimum Inverter Capacity Design for Three-Phase Four-Wire LC-HAPF

Owing to the limitations among the existing literatures [19–26], there is still no mathematical deduction for the design of the LC-HAPF minimum dc-link voltage in current harmonics and reactive power compensation. Therefore, the key contribution of this section is to investigate and discuss the minimum dc-link voltage and also inverter capacity design for the three-phase four-wire LC-HAPF.

In this section, the equivalent circuit model in a-b-c coordinate of the three-phase four-wire center-split LC-HAPF as shown in Fig. 4.5 is initially introduced. According to the current quality of the loading and the LC-HAPF single-phase equivalent circuit models, the minimum dc-link voltage expression for the LC-HAPF in current harmonics and reactive power compensation is proposed. Finally, representative simulation and experimental results of the LC-HAPF are given to verify its effectiveness of the minimum dc-link voltage design expression. Given that most of the loadings in the distribution power systems are inductive, the following analysis and discussion only focus on inductive nonlinear loads [39].

4.7.1 Mathematical Modeling of a Three-Phase Four-Wire Center-Split LC-HAPF in A-B-C Coordinate

The equivalent circuit model of LC-HAPF in a-b-c coordinate of the three-phase four-wire center-split LC-HAPF as shown in Fig. 4.5 is shown in Fig. 4.54, where the subscript 'x' denotes phase $x = a, b, c, n$. v_x is the load voltage, i_{sx}, i_{Lx} and i_{cx} are the system, load, and inverter current for each phase respectively. C_{c1} and L_{c1} are the coupling capacitor and inductor of the LC-HAPF. v_{Cc1x} and v_{Lc1x} are the coupling capacitor voltage and inductor voltage. v_{inv1x} is the inverter output voltage.

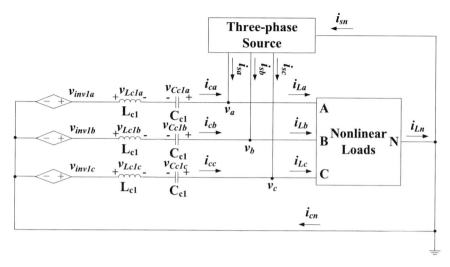

Fig. 4.54 Equivalent circuit model in *a-b-c* coordinates of the three-phase four-wire center-split LC-HAPF

From Fig. 4.54, the differential equations of the coupling inductor L_{c1} and capacitor C_{c1} can be expressed as:

$$L_{c1} \frac{d}{dt} \begin{bmatrix} i_{ca} \\ i_{cb} \\ i_{cc} \end{bmatrix} = \begin{bmatrix} v_{inv1a} \\ v_{inv1b} \\ v_{inv1c} \end{bmatrix} - \left(\begin{bmatrix} v_a \\ v_b \\ v_c \end{bmatrix} - \begin{bmatrix} v_{Cc1a} \\ v_{Cc1b} \\ v_{Cc1c} \end{bmatrix} \right) \tag{4.64}$$

$$C_{c1} \frac{d}{dt} \begin{bmatrix} v_{Cc1a} \\ v_{Cc1b} \\ v_{Cc1c} \end{bmatrix} = - \begin{bmatrix} i_{ca} \\ i_{cb} \\ i_{cc} \end{bmatrix} \tag{4.65}$$

In the following, the minimum dc-link voltage expressions for the LC-HAPF will be presented.

4.7.2 Minimum Inverter Capacity Analysis of a Three-Phase Four-Wire Center-Split LC-HAPF

From the previous analysis results, Fig. 4.55 shows the LC-HAPF single-phase equivalent circuit models in *a-b-c* coordinate. In the following analysis, all ac parameters are in RMS value except dc components. From Fig. 4.55, the required inverter fundamental output voltages (V_{inv1xf}) and inverter harmonic output voltages (V_{inv1xn}) at each harmonic order can be found, where the subscripts '*f*' and '*n*' denote the fundamental and harmonic frequency components. As V_{inv1xf} and V_{inv1xn}

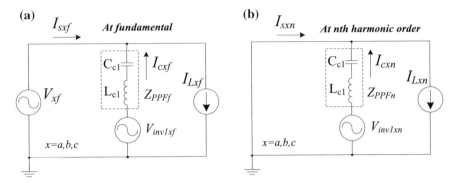

Fig. 4.55 LC-HAPF single-phase equivalent circuit models in *a-b-c* coordinate: **a** at fundamental frequency and **b** at *n*th harmonic order frequency

are in RMS values, the minimum dc-link voltage values (V_{dc1xf} and V_{dc1xn}) for compensating the phase fundamental reactive current component and each *n*th harmonic current component are calculated as the peak values of the required inverter fundamental and harmonic output voltages, in which $V_{dc1xf} = \sqrt{2}V_{inv1xf}$ and $V_{dc1xn} = \sqrt{2}V_{inv1xn}$ respectively. In order to provide sufficient dc-link voltage for compensating load reactive and harmonic currents, the minimum dc-link voltage requirement (V_{dc1x}) of the LC-HAPF single-phase circuit model as shown in (4.66) is deduced by considering the worst phase relation between each harmonic component, in which their corresponding peak voltages of the VSI at AC side are assumed to be superimposed.

$$V_{dc1x} = \sqrt{\left|V_{dc1xf}\right|^2 + \sum_{n=2}^{\infty}\left|V_{dc1xn}\right|^2} \qquad (4.66)$$

From Fig. 4.55a, when the load voltage V_x is pure sinusoidal without harmonic components, $V_x = V_{xf}$, the inverter fundamental output voltage of the LC-HAPF single-phase circuit model can be expressed as:

$$V_{inv1xf} = V_x + Z_{PPFf} \times I_{cxf} \qquad (4.67)$$

Since most of the loadings in the distribution power systems are inductive, the fundamental impedance of the coupling capacitor C_{c1} should be larger than that of the coupling inductor L_{c1} which yields $Z_{PPFf} = \left|Z_{PPFf}\right|e^{j\varphi f}$, $\varphi f = -90°$ When the LC-HAPF operates at the ideal case, the fundamental compensating current I_{cxf} contains the pure reactive component I_{cxfq} only without the active current component I_{cxfp}, therefore (4.67) can be rewritten as:

$$V_{inv1xf} = V_x - \left|\omega L_{c1} - \frac{1}{\omega C_{c1}}\right| |I_{cxfq}|, \tag{4.68}$$

where $\omega = 2\pi f$.

From Fig. 4.55b, the inverter harmonic output voltage V_{inv1xn} at nth ($n = 1$, 2 ... ∞) harmonic order can be expressed as:

$$V_{inv1xn} = \left|n\omega L_{c1} - \frac{1}{n\omega C_{c1}}\right| |I_{cxn}| \quad n = 2, 3...\infty, \tag{4.69}$$

where I_{cxn} is the nth order harmonic compensating current.

From Fig. 4.5, when the LC-HAPF compensates power quality issues, the absolute reactive and nth order harmonic compensating current should be equal to those of the loading, it yields:

$$|I_{cxfq}| = |I_{Lxfq}|, |I_{cxn}| = |I_{Lxn}|, \tag{4.70}$$

where I_{Lxfq} and I_{Lxn} are the reactive and nth order harmonic current of the loading.

From (4.67) to (4.70), the inverter fundamental and each nth harmonic order output voltages of the LC-HAPF single-phase circuit model (V_{inv1xf} and V_{inv1xn}) can be calculated. Then the minimum dc-link voltage requirement (V_{dc1x}) for the LC-HAPF single-phase circuit model can be found by (4.66).

By using the generalized single-phase p-q theory [29], the reactive power and current harmonics in each phase can be compensated independently, thus the final required minimum dc-link voltage for the three-phase four-wire center-split LC-HAPF (V_{dc1}) will be the maximum one among the calculated minimum values of each phase (V_{dc1x}), which can be expressed as:

$$V_{dc1} = \max(2V_{dc1a}, 2V_{dc1b}, 2V_{dc1c}) \tag{4.71}$$

Thus, the inverter capacity of the LC-HAPF (S_{inv}) can be expressed as:

$$S_{inv} = 3\frac{V_{dc1}}{\sqrt{2}}I_c \tag{4.72}$$

where $I_c = \max(I_{ca} = I_{cb} = I_{cc})$. From (4.72), the inverter capacity of the LC-HAPF is proportional to the corresponding dc-link voltage. Thus, the dc-link voltage level can reflect the inverter capacity of the LC-HAPF. Table 4.24 summarizes the minimum dc-link voltage deduction steps of the LC-HAPF.

From Table 4.24, the minimum dc-link voltage value of the LC-HAPF can be calculated only when the spectra of the load currents are known. If the load current spectra cannot be measured (i.e. unknown loads) before the installation of the LC-HAPF, via Fast Fourier Transform (FFT), the load current spectra can be figured out by the DSP of the LC-HAPF after installation. Then the minimum dc-link voltage value can be calculated by the DSP through the deduction steps in

Table 4.24 Minimum dc-link voltage deduction steps of the LC-HAPF

1	Inverter fundamental output voltage	
	$V_{inv1xf} = V_x - \left|\omega L_{c1} - \frac{1}{\omega C_{c1}}\right|\left|I_{cxfq}\right|$	(4.68)
	where $\left|I_{cxfq}\right| = \left|I_{Lxfq}\right|$, $\omega = 2\pi f$	(4.70)
2	Inverter nth harmonic order output voltage	
	$V_{inv1xn} = \left|n\omega L_{c1} - \frac{1}{n\omega C_{c1}}\right|\left|I_{cxn}\right|$	(4.69)
	where $\left|I_{cxn}\right| = \left|I_{Lxn}\right|$, $n = 2, 3 \ldots \infty$, $\omega = 2\pi f$	(4.70)
3	Minimum dc-link voltage	
	$V_{dc1} = \max(2V_{dc1a}, 2V_{dc1b}, 2V_{dc1c})$	(4.71)
	Where $V_{dc1x} = \sqrt{\left|V_{dc1xf}\right|^2 + \sum_{n=2}^{\infty} \left|V_{dc1xn}\right|^2}$	(4.66)
	$V_{dc1xf} = \sqrt{2}V_{inv1xf}$, $V_{dc1xn} = \sqrt{2}V_{inv1xn}$, where $n = 2, 3 \ldots \infty$	

Table 4.24. Once the minimum dc-link voltage value is known, the LC-HAPF system can start operation. Through the dc-link voltage control method, its dc-link voltage can be controlled to reach this minimum reference value.

4.7.3   Simulation and Experimental Verifications for Minimum Inverter Capacity Analysis of the Three-Phase Four-Wire LC-HAPF

To verify the minimum dc-link voltage analysis, representative simulation and experimental results of the three-phase four-wire center-split LC-HAPF system as shown in Fig. 4.5 will be given. In order to simplify the verification, the dc-link is supported by external dc voltage source and the simulated and experimental three-phase loadings are approximately balanced. Table 4.25 lists the simulated and experimental system parameters for the LC-HAPF. From Table 4.25, the coupling

Table 4.25 LC-HAPF system parameters for simulations and experiments

System parameters		Physical values
System source-side	V_x	220 V
	f	50 Hz
LC-HAPF	L_{c1}	8 mH
	C_{c1}	50 μF
	V_{dc1}	45, 65, 90 V
Nonlinear rectifier load	R_{NL1x}	43.2 Ω
	L_{NL1x}	34.5 mH
	C_{NL1x}	392.0 μF

L_{c1} and C_{c1} are designed based on the load fundamental reactive power consumption and tuned at the 5th order harmonic frequency. As the simulated and experimental loadings are approximately balanced, only phase a compensation diagrams will be illustrated.

Simulation studies were carried out by using Matlab. In order to verify the simulation results, a 220 V/10 kVA three-phase four-wire center-split LC-HAPF experimental prototype is implemented. IGBTs are employed as the switching devices for the active inverter part. And the control system of this prototype is a DSP TMS320F2812 and the analog to digital (A/D) sampling frequency of the LC-HAPF system is set as 40 μs (25 kHz). Figure 4.6 shows the reactive and harmonic reference compensating current deduction and PWM control block diagram for the three-phase four-wire LC-HAPF system. Hysteresis PWM is applied for generating the required compensating current. In the following, since the load harmonic current contents beyond the 9th order are small, for simplicity, the required minimum dc-link voltage calculation will be taken into account up to 9th harmonic order only.

4.7.3.1 Simulation Results

Figure 4.56a illustrates the simulated load current i_{Lx} waveform and its spectrum of phase a, in which its corresponding fundamental reactive current, 3rd, 5th, 7th and 9th order harmonic current in RMS values are shown in Table 4.26. From Fig. 4.56a and Table 4.28, the total harmonic distortion of the load current (THD_{iLx}) is 32.1 % and the load neutral current (i_{Ln}) is 5.35 A, in which the 3rd and 5th order harmonic contents dominate the load harmonic current. With the help of Tables 4.24 and 4.27 shows the required minimum dc-link voltage values (V_{dc1xf} and V_{dc1xn}) for compensating the fundamental reactive current, 3rd, 5th, 7th and 9th harmonic current components and the minimum dc-link voltage (V_{dc1}) of the LC-HAPF, in which $V_{dc1} = 79.24$ V. The dc-link voltage for the LC-HAPF is chosen as $V_{dc1} = 45$ V, 65 V, 90 V for performing compensation respectively. After the compensation by the LC-HAPF, Fig. 4.56b–d show the simulated system current i_{sx} waveforms and their spectra of phase a at different dc-link voltage levels. Their corresponding results are summarized in Table 4.28.

Table 4.26 Simulated and experimental fundamental reactive current, 3rd, 5th, 7th and 9th orders harmonic current values of the loading

	Fundamental reactive current (A)	3rd order harmonic current (A)	5th order harmonic current (A)	7th order harmonic current (A)	9th order harmonic current (A)
Simulation results	3.72	1.96	0.53	0.23	0.16
Experimental results	3.41	1.92	0.45	0.20	0.12

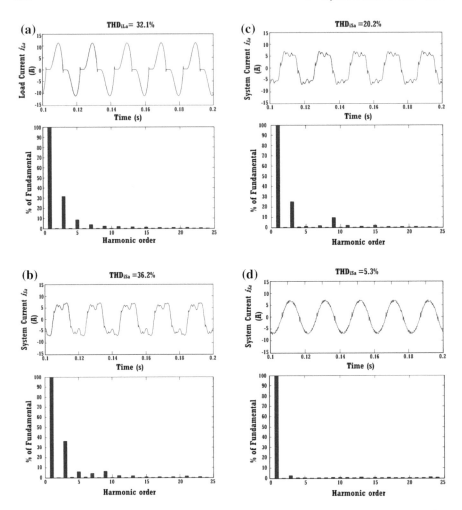

Fig. 4.56 Simulated i_{sx} and its spectrum of phase a before and after LC-HAPF compensation: **a** before compensation, **b** after compensation with $V_{dc1} = 45$ V, **c** after compensation with $V_{dc1} = 65$ V and **d** after compensation with $V_{dc1} = 90$ V

Table 4.27 Simulated and experimental required dc-link voltage of the LC-HAPF

LC-HAPF	V_{dc1xf} (V)	V_{dc1x3} (V)	V_{dc1x5} (V)	V_{dc1x7} (V)	V_{dc1x9} (V)	V_{dc1} (V)
Simulation results	10.56	37.92	0.13	2.77	3.52	79.24
Experimental results	16.42	37.15	0.10	2.40	2.64	81.54

Table 4.28 Summary of simulated and experimental results before and after LC-HAPF compensation

LC-HAPF		V_{dc1} (V)	3rd harmonic (%)	5th harmonic (%)	DPF	THD_{isx} (%)	i_{sn} (A)
Simulation results	Without comp.	–	31.5	8.6	0.80	32.1	5.35
	After comp.	45	36.2	6.1	1.00	36.2	5.85
		65	25.2	1.1	1.00	20.2	3.60
		90	2.7	0.6	1.00	5.3	0.86
Experimental results	Without comp.	–	31.4	7.3	0.83	32.7	5.77
	After comp.	45	23.8	3.2	1.00	25.3	3.51
		65	14.8	2.4	1.00	17.2	2.45
		90	6.3	1.1	1.00	8.0	1.30

From Tables 4.27 and 4.28, when the LC-HAPF is operating, the coupling LC mainly eliminates the 5th order harmonic current. From Fig. 4.56b–d, with the dc-link voltage of $V_{dc1} = 45$ V ($<V_{dc1} = 79.24$ V), the LC-HAPF cannot perform current compensation effectively. After compensation, the total harmonic distortion of phase a system current is $THD_{isx} = 36.2$ % and the system neutral current (i_{sn}) is 5.85 A, in which the compensated THD_{isx} does not satisfy the international standards [31, 33]. When the dc-link voltage increases to $V_{dc1} = 65$ V, because this value is closer to the required $V_{dc1} = 79.24$ V, the LC-HAPF can obtain better compensating performances with $THD_{isx} = 20.2$ % and $i_{sn} = 3.60$ A, in which the compensated THD_{isx} still does not satisfy the international standards [31, 33]. When the dc-link voltage increases to $V_{dc1} = 90$ V, the LC-HAPF then can effectively track the reference compensating current and achieve the best compensating performance with $THD_{isx} = 5.3$ % and $i_{sn} = 0.86$ A among the three cases.

4.7.3.2 Experimental Results

Figure 4.57a illustrates the experimental i_{Lx} waveform and its spectrum of phase a, in which its corresponding fundamental reactive current, the 3rd, 5th, 7th and 9th order harmonic currents in RMS value of the loading are shown in Table 4.26. From Fig. 4.57a and Table 4.28, the THD_{iLx} is 32.7 % and the i_{Ln} is 5.77 A, in which the 3rd and 5th orders harmonic contents dominate the load harmonic current. With the help of Tables 4.24 and 4.27 shows the required V_{dc1xf} and V_{dc1xn}, in which $V_{dc1} = 81.54$ V. Similar as simulation part, the dc-link voltage for the LC-HAPF is chosen in three levels as $V_{dc1} = 45$, 65, and 90 V for performing compensation respectively. After compensation by the LC-HAPF, Fig. 4.57b–d show the experimental i_{sx} waveforms and their spectra of phase a at different dc-link voltage levels. Their corresponding experimental results are summarized in Table 4.28.

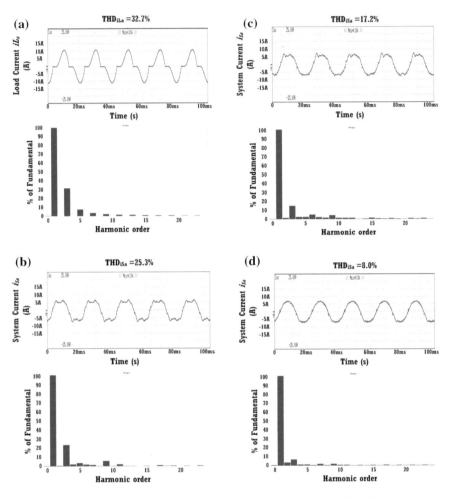

Fig. 4.57 Experimental i_{sx} and its spectrum of phase a before and after LC-HAPF compensation: **a** before compensation, **b** after compensation with $V_{dc1} = 45$ V, **c** after compensation with $V_{dc1} = 65$ V and **d** after compensation with $V_{dc1} = 90$ V

From Tables 4.27 and 4.28, when the LC-HAPF operates, the coupling LC mainly eliminates the 5th order harmonic current. From Fig. 4.57b–d, with $V_{dc1} = 45$ V ($<V_{dc1} = 81.54$ V), the LC-HAPF cannot perform current compensation effectively. After compensation, the THD_{isx} of phase a is 25.3 % and the i_{sn} is 3.51 A, in which the compensated THD_{isx} does not satisfy the international standard [31, 33]. When the dc-link voltage increases to $V_{dc1} = 65$ V, because this value is closer to the required $V_{dc1} = 81.54$ V, the LC-HAPF can obtain better compensating performances with $THD_{isx} = 17.2$ % and $i_{sn} = 2.45$ A, yet the compensated THD_{isx} does not satisfy the international standard [31]. When the dc-link voltage

increases to V_{dc1} = 90 V, the LC-HAPF can achieve the best compensating performance with THD_{isx} = 8.0 % and i_{sn} = 1.30 A among the three cases.

The experimental results are similar to the simulated results, which verify the LC-HAPF minimum dc-link voltage expression. To obtain LC-HAPF good compensating performances, its dc-link operating voltage can be chosen as the calculated V_{dc1}. Then its minimum inverter capacity can also be found by (4.72).

4.8 Design and Performance of a 220 V/10 kVA Three-Phase Four-Wire LC-HAPF Experimental Prototype

4.8.1 System Configuration of Three-Phase Four-Wire LC-HAPF

The system configuration of the 220 V/10 kVA three-phase four-wire LC-HAPF experimental prototype is shown in Fig. 4.58 while the LC-HAPF experimental prototype testing environment is shown in Fig. 4.59. From Fig. 4.58, the subscript 'x' denotes phase a, b, c, n. v_x is the load voltage, i_{sx}, i_{Lx} and i_{cx} are the system, load and inverter currents for each phase. C_{c1} and L_{c1} are the coupling capacitor and

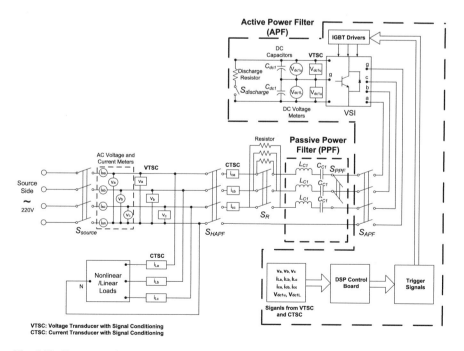

Fig. 4.58 System configuration of the 220 V/10 kVA LC-HAPF experimental prototype

Fig. 4.59 The
220 V/10 kVA LC-HAPF
experimental prototype
testing environment

Table 4.29 A 220 V/10 kVA LC-HAPF experimental system parameters

System parameters		Physical values
System source-side	V_x	220 V
	f	50 Hz
Coupling LC ($Q_{cxf\_PPF} \approx -790.0$ VAR)	L_{c1}	8 mH
	C_{c1}	50 μF
DC capacitor	C_{dc1}	3.3 mF
DC-link voltage levels	V_{dc1u}, V_{dc1L}	25, 50, 75 V

inductor, L_{cn} is the coupling neutral inductor. C_{dc1}, V_{dc1U} and V_{dc1L} are the dc capacitor, upper and lower dc capacitor voltages with $V_{dc1U} = V_{dc1L} = 0.5V_{dc1}$. S_{source}, S_{HAPF}, S_{APF}, S_{PPF}, S_R, $S_{discharge}$ are controllable switches for controlling different parts of the circuit. VTSC and CTSC represent voltage and current transducers with signal conditioning board. The load is a nonlinear load, a linear load or their combination. Table 4.29 shows the system parameters of the 220 V/10kVA LC-HAPF experimental prototype, and three adaptive dc-link voltage levels (V_{dc1u}, V_{dc1L} = 25, 50 and 75 V) are designed.

4.8.2 Balanced and Unbalanced Testing Loads

The 220 V/10 kVA three-phase four-wire center-split LC-HAPF experimental prototype has been tested under balanced and unbalanced loading situations as shown in Figs. 4.60 and 4.61. Their corresponding parameters' values are summarized in Table 4.30.

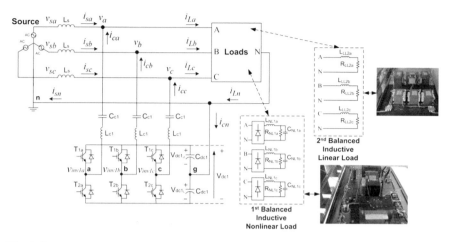

Fig. 4.60 The 220 V/10 kVA LC-HAPF experimental prototype under balanced loading situation

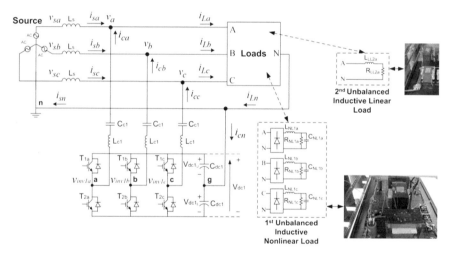

Fig. 4.61 The 220 V/10 kVA LC-HAPF experimental prototype under unbalanced loading situation

4.8.3 Design of Coupling LC of LC-HAPF

For the full bridge diode rectifier loading as shown in Figs. 4.60 and 4.61, the 3rd and 5th order harmonic currents will be the two dominant harmonic current contents. For designing the coupling LC parameters based on the average fundamental reactive power consumption $\bar{Q}_{Lxf} \approx 790.0$ var and $n_1 = 5$, from (4.1)–(4.3), the coupling capacitor and inductor can be determined as $C_{c1} \approx 50 \ \mu F$, $L_{c1} \approx 8 \ mH$

Fig. 4.62 Hardware circuit
diagram of coupling LC

respectively. The coupling passive part (C_{c1}, L_{c1}) hardware circuit diagram of the
220 V/10 kVA LC-HAPF experimental system is shown in Fig. 4.62.

4.8.4   Design of Active Inverter Part of LC-HAPF

The active inverter part of the LC-HAPF is composed of a DC/AC center-split
voltage source inverter (VSI) with dc-link capacitors, which includes IGBT power
switches with drivers, transducers with signal conditioning boards and digital
controller. The design of the following three components of the APF part will be
presented as follows:

(1)   IGBT power switches with drivers,
(2)   Transducer with signal conditioning boards,
(3)   Digital controller and its software design.

4.8.4.1   IGBT Power Switches with Drivers

Presently   available   power   semiconductor   switches   include   bipolar
technology-based devices such as bipolar junction transistors (BJTs), thyristors and
gate turn-off (GTO) thyristors, and metal-oxide-semiconductor (MOS)-based
devices such as MOS field effect transistors (MOSFETs), insulated gate bipolar
transistors (IGBTs), and MOS-controlled thyristors (MCTs). A summary of power
device capabilities is shown in Fig. 4.63 [44].

The Mitsubishi third generation IGBT PM300DSA60 dual intelligent power
module (IPM) is selected as the power switching devices for the 220 V/10 kVA
LC-HAPF experimental system, with maximum rated current and voltage of 300 A
and 600 V respectively [45]. The appearance and circuit diagram of the selected
dual IPM is shown in Chap. 3, in which this module can be used as one leg of the
VSI, which provides great convenience to the hardware implementation.

The schematic diagram of the IGBT driver board is shown in Fig. 4.64, which is
used to drive one dual IPM. The IGBT driver board is designed to control the

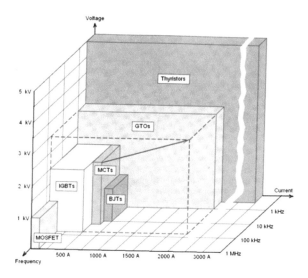

Fig. 4.63 Summary of power semiconductor device capabilities [44]

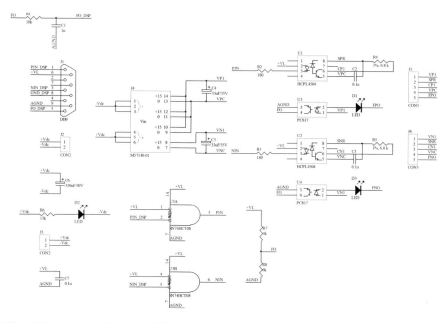

Fig. 4.64 Schematic diagram of IGBT driver board

switching state of IGBT with the trigger signals generated from the digital controller. The final hardware connection between the IGBT power switches and drivers for the 220 V/10 kVA LC-HAPF experimental system is shown in Fig. 4.65. The key design consideration of the IGBT driver board has been discussed in Chap. 3.

Fig. 4.65 Hardware connection between IGBT power switches and drivers for the 220 V/10 kVA experimental prototype

4.8.4.2 Transducers with Signal Conditioning Boards

The three-phase load voltages, load currents, compensating currents and dc-link voltages of the LC-HAPF are measured by transducers with signal conditioning boards. The adopted transducers are based on the Hall-Effect transducer, which provides an isolated measurement for dc and ac voltage and current. The voltage and current signal conditioning boards can transfer the large electrical signals into small analog signals in order to be adopted as the A/D converter inputs of the digital control system. The measured output signals from the signal conditioning boards are sent to the A/D converter and converted into digital signals in the digital controller. These signals are required to determine the reference compensating currents. The schematic diagram of the signal conditioning circuit is shown in Fig. 4.66. From Fig. 4.66, the signal conditioning circuit mainly includes one voltage follower and two negative gain amplifiers. The voltage follower can provide a high input impedance to avoid loading effect, while the two negative gain amplifiers can provide an appropriate positive output electrical signal to A/D input of the digital controller. A 3.3 V zener diode is implemented at the output side to limit the input voltage level to a 3.3 V digital controller.

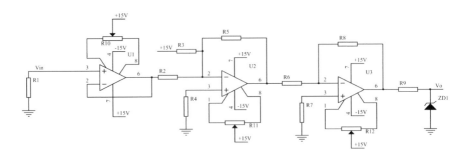

Fig. 4.66 Schematic diagram of the signal condition circuit

In order to synthesize parameters of the signal conditioning circuit, the R1 of KT 20 A/P (current transducer) indicated in Fig. 4.66 is selected as 50 Ω, while that of KV 50 A/P (voltage transducer) is selected as 100 Ω [46]. Therefore, the current and voltage transducer output voltage (V_{in}) will have the same range between $-5V_{peak}$ to $5V_{peak}$.

The selected A/D converter of the digital controller has an input range of 0–3.3 V. Therefore, the output voltage (V_o) of the signal conditioning circuits should be consistent with (0–3.3 V). To avoid the unsteady state when the input voltage approaches 3.3 V, the peak output voltage is changed to the range of 0.3–3.0 V. According to the schematic diagram as shown in Fig. 4.66, the signal conditioning board output voltage can be expressed as:

$$V_o = \frac{R_5}{R_2} V_{in} + \frac{R_5}{R_3} \times 15 \tag{4.73}$$

For ac measurement, a dc offset should be provided for the input analog signals because a negative voltage is not compatible for a 0–3.3 V digital controller. Set R5 = 2.2 kΩ and R3 = 20 kΩ, then $V_o = \frac{2.2k}{R_2} V_{in} + 1.65$.

According to different experimental loading conditions as listed in Tables 4.30, 4.32 and 4.33, the maximum measurement ranges for load currents and compensating currents are set as $\pm 20A_{peak}$ and $\pm 20A_{peak}$ respectively while the maximum measurement range of the load voltages and dc-link capacitor voltages are set as $\pm 330 V_{peak}$ and $+200V_{peak}$ respectively. Table 4.31 shows the design parameters of the signal conditioning boards. The photo of the overall current and voltage transducer with signal condition boards for the 220 V/10 kVA LC-HAPF experimental prototype is shown in Fig. 4.67.

Table 4.30 Experimental parameters for balanced and unbalanced loading situations

System parameters					Physical values
V_x, f					220 V, 50 Hz
Balanced loading	1st inductive nonlinear load	A	$R_{NL1x}, L_{NL1x},$ C_{NL1x}		43.2 Ω, 34.5 mH, 392.0 μF ($Q_{Lxf} \approx 720.0$ var)
		B			
		C			
	2nd inductive linear load	A	R_{LL2x}, L_{LL2x}		60 Ω, 70 mH ($Q_{Lxf} \approx 200.0$ var)
		B			
		C			
Unbalanced loading	1st inductive nonlinear load	A	$R_{NL1x}, L_{NL1x},$ C_{NL1x}		48.2 Ω, 34.5 mH, 392.0 μF ($Q_{Lxf} \approx 660.0$ var)
		B			43.2 Ω, 34.5 mH, 392.0 μF ($Q_{Lxf} \approx 720.0$ var)
		C			
	2nd inductive linear load	A	R_{LL2x}, L_{LL2x}		80 Ω, 70 mH ($Q_{Lxf} \approx 100.0$ var)

Table 4.31 Parameters of the signal conditioning boards (in peak value)

Max. measurement range (A_{peak}/V_{peak})		V_{in} (V_{peak})	V_o (V_{peak})	R_2 (kΩ)	R_3 (kΩ)	R_5 (kΩ)
Current	i_{Lx}	−20–20	−5–5	0.309–2.991	8.2	20	2.2
	i_{cx}	−20–20	−5–5	0.309–2.991	8.2	20	2.2
Voltage	v_x	−330–330	−5–5	0.309–2.991	8.2	20	2.2
	V_{dc1U}, V_{dc1L}	0–200	0–5	0.330–2.991	8.2	100	2.2

Note R4 = 1.5 kΩ, R7 = 5 kΩ and R6 = R8 = 10 kΩ, R10, R11, R12 are 20 kΩ adjustable resistor

Table 4.32 Experimental results before LC-HAPF compensation under balanced loading case

Before LC-HAPF compensation under balanced loading case							
Different cases		Q_{sxf} (var)	PF	THD_{isx} (%)	THD_{vx} (%)	i_{sx} (A)	i_{sn} (A)
1st inductive loading	A	723.0	0.804	32.5	1.7	6.506	5.808
	B	718.5	0.805	31.5	1.9	6.467	
	C	721.2	0.804	31.6	1.5	6.444	
1st and 2nd inductive loading	A	921.3	0.870	21.3	1.6	9.520	5.659
	B	920.1	0.872	20.5	1.8	9.539	
	C	921.9	0.870	20.7	1.5	9.493	

Table 4.33 Experimental results before LC-HAPF compensation under unbalanced loading case

Before LC-HAPF compensation under unbalanced loading case							
Different cases		Q_{sxf} (var)	PF	THD_{isx} (%)	THD_{vx} (%)	i_{sx} (A)	i_{sn} (A)
1st inductive loading	A	665.5	0.808	33.9	1.6	6.072	5.750
	B	719.5	0.807	31.5	1.9	6.468	
	C	721.8	0.802	31.6	1.7	6.457	
1st and 2nd inductive loading	A	765.5	0.877	23.2	1.6	8.357	6.067
	B	719.5	0.807	31.5	1.9	6.468	
	C	721.8	0.802	31.6	1.7	6.457	

Fig. 4.67 Overall current and voltage transducers with signal conditioning boards for the 220 V/10 kVA LC-HAPF experimental prototype

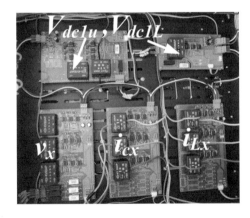

4.8.4.3   Software Design of Digital Controller DSP-TMS320F2812

The reference compensating current calculation is achieved by a digital signal processor (DSP). The high speed DSP-TMS320F2812 is chosen which has high performance in the real-time control and motor/machine control. There are total sixteen multiplexed analog inputs of 12-bit analog-to-digital converter (ADC) core with built-in sample and hold (S/H) in the DSP-TMS320F2812. The 12-bit ADC module is operating with a fast conversion rate of 80 ns at 25 MHz ADC clock. The ADC sequencer consists of two independent 8-state sequencers that can also be cascaded together to form one 16-state sequencer [47]. After the conversion, the digital value of the selected channel is stored in the appropriate result register. The results will be transformed into the corresponding format and values by bit shift operation on the result registers.

There are two Event Managers (EV), namely EVA and EVB, embedded in TMS320F2812. Each EV module contains two general-purpose (GP) timers. The GP timers can be used to activate the A/D conversion to provide a time base for the operation of the full compare units, or to calculate the reference current. There are six PWM outputs with programmable deadband and output logic for each EV created by the intrinsic three full compare units. These PWM signals are generated by using EV module, Timer 1 (EVA) and Timer 3 (EVB). The deadband unit is used to implement the deadtime of the inverter in each of the compare units. When the deadband unit is enabled for the comparing units, the transition edges of the two signals are separated by a time interval determined by the inverter switch limitation. A deadband can be provided to avoid the short-circuit case between the upper and lower switches in one leg. The deadband is designed as 4.27 µs, which is larger than the recommended condition (≥ 3.5 µs) of the PM300DSA60 IPM.

For the 220 V/10 kVA LC-HAPF experimental prototype, Timer 2 (EVA) is used to define the sample rate of ADC module, in which the sampling frequency is set to 25 kHz. For every 1/25 kHz(s) period, Timer 2 will provide a trigger signal to process AD conversion and the corresponding interrupt. There are totally 4 groups with 12 channel signals, i.e. 3 load voltages v_x, 3 load currents i_{Lx}, 3 compensating currents i_{cx}, and 2 dc-link capacitor voltages V_{dc1U}, V_{dc1L} are converted into digital values. Timer 1 (EVA) is responsible for generating PWM. The frequency of Timer 1 is determined by the switching frequency limitation of the IGBT. Timer 1 is set to have a maximum switching frequency of 12.5 kH. There is only one interrupt activated, in which the A/D interrupt has the highest priority. In the following, the program flow chart for the DSP-TMS320F2812 in performing ADC signal sampling, single-phase instantaneous p-q compensation algorithm, adaptive dc-link voltage control, and generating the PWM signals for controlling the VSI output is shown in Fig. 4.68.

The DSP controller for the LC-HAPF experimental prototype is shown in Fig. 4.69, in which there is a signals connection printed circuit board (PCB) for convenient connection between the peripherals I/O ports of DSP and other physical devices.

(a)   **(b)**

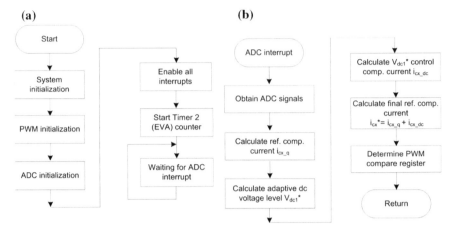

Fig. 4.68 DSP program flowchart of the LC-HAPF: **a** main program, **b** interrupt service routine

Fig. 4.69 The DSP
controller for the
220 V/10 kVA LC-HAPF
experimental prototype

4.8.5 Experimental Results for a 220 V/10 kVA Three-Phase Four-Wire LC-HAPF Experimental Prototype

4.8.5.1 Power Quality Data of the Experimental Balanced and Unbalanced Loadings

Before the LC-HAPF performs compensation, Figs. 4.70 and 4.71 show the experimental reactive power at system source-side Q_{sxf}, load voltage v_x and system current i_{sx} waveforms for both balanced and unbalanced loading situations. Tables 4.32 and 4.33 summarize the power quality parameters for both balanced and unbalanced loading situations.

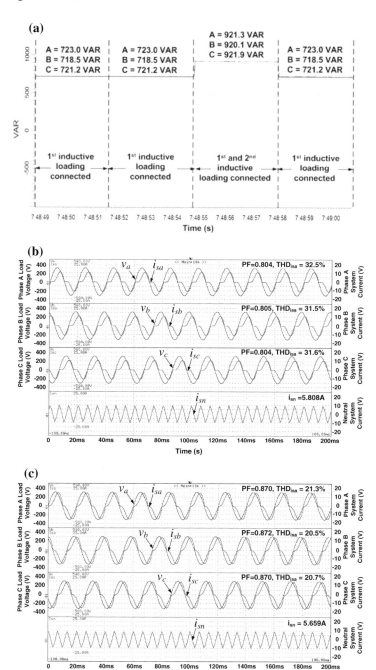

Fig. 4.70 Before LC-HAPF compensation under balanced loading case: **a** Q_{sxf}, **b** v_x and i_{sx} when the 1st loading is connected, and **c** v_x and i_{sx} when the 1st and 2nd loading are connected

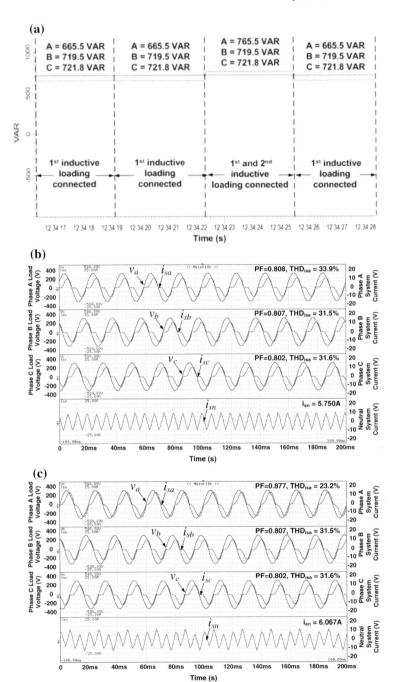

Fig. 4.71 Before LC-HAPF compensation under unbalanced loading case: **a** Q_{sxf}, **b** v_x and i_{sx} when the 1st loading is connected, and **c** v_x and i_{sx} when the 1st and 2nd loading are connected

Considering the balanced loading case, when the 1st inductive loading is connected, the three-phase Q_{sxf} are 723.0, 718.5 and 721.2 var with power factor (PF) = 0.804, 0.805 and 0.804 respectively, and the total harmonic distortion (THD_{isx}) of i_{sx} are 32.5, 31.5 and 31.6 %, in which the THD_{isx} does not satisfy the international standards [31, 33]. When both the 1st and 2nd inductive loadings are connected, the three-phase Q_{sxf} will increase to 921.3, 920.1 and 921.9 var with PF = 0.870, 0.872 and 0.870 respectively, then the THD_{isx} become 21.3, 20.5 and 20.7 %, in which the THD_{isx} does not satisfy the international standards [31, 33].

Considering the unbalanced loading case, when the 1st inductive loading is connected, the three-phase Q_{sxf} are 665.5, 719.5 and 721.8 var with PF = 0.808, 0.807 and 0.802 respectively, and the THD_{isx} of i_{sx} are 33.9, 31.5 and 31.6 %, in which the THD_{isx} does not satisfy the international standards [31, 33]. When both the 1st and 2nd inductive loadings are connected, phase A Q_{sxf} will increase to 765.5 var with PF = 0.877, and the THD_{isx} of phase A becomes 23.2 %, in which the three-phase THD_{isx} does not satisfy the international standards [31, 33].

As the experimental load harmonic current contents beyond the 9th order are small, for simplicity, the LC-HAPF required minimum dc-link voltage for current harmonic compensation will be taken into account up to the 9th harmonic order only. Table 4.34 shows the 3rd, 5th, 7th and 9th order harmonic current in RMS values under balanced and unbalanced loading situations.

With the help of the analysis results from Sects. 4.6 and 4.7, based on Tables 4.32, 4.33 and 4.34, the LC-HAPF final required minimum dc-link voltage (V_{dc1u}, V_{dc1L}) for both reactive power and current harmonics compensation can be calculated. And the chosen adaptive voltage levels are illustrated in Table 4.35.

Under the balanced and unbalanced loading situations as shown in Figs. 4.70, 4.71 and Tables 4.32, 4.33, the corresponding compensation results by the 220 V/10 kVA LC-HAPF experimental prototype with conventional fixed dc-link voltage control and adaptive dc-link voltage control are presented in the following.

Table 4.34 Experimental 3rd, 5th, 7th and 9th orders load harmonic current values under balanced and unbalanced loading cases

Different situations			Harmonic current order			
			3rd order (A)	5th order (A)	7th order (A)	9th order (A)
Balanced loading	1st loading	A, B, C	1.92	0.45	0.20	0.12
	1st and 2nd loadings	A, B, C	1.90	0.46	0.23	0.12
Unbalanced loading	1st loading	A	1.86	0.39	0.18	0.11
		B, C	1.92	0.45	0.20	0.12
	1st and 2nd loadings	A	1.81	0.40	0.20	0.11
		B, C	1.92	0.45	0.20	0.12

Table 4.35 LC-HAPF experimental minimum dc-link voltage under balanced and unbalanced loading cases

Different situations			Chosen adaptive voltage levels (V_{dc1u}, V_{dc1L}) (V)
Balanced loading	1st loading	A, B, C	50
	1st and 2nd loadings	A, B, C	75
Unbalanced loading	1st loading	A	75
		B, C	
	1st and 2nd loadings	A	50
		B, C	

4.8.5.2 Experimental Results of Fixed DC-Link Voltage Control

In the balanced loading situations as shown in Fig. 4.70 and Table 4.32, with a fixed dc-link voltage reference of V_{dc1u}, V_{dc1L} = 75 V, Fig. 4.72 shows the whole experimental dynamic compensation process with a fixed dc-link voltage control scheme. Figure 4.72 includes the waveforms of: (1) V_{dc1u}, V_{dc1L}, (2) Q_{sxf}, (3) v_x and i_{sx} of phase a after LC-HAPF starts operation and (4) v_x and i_{sx} of phase a after the 2nd loading is connected.

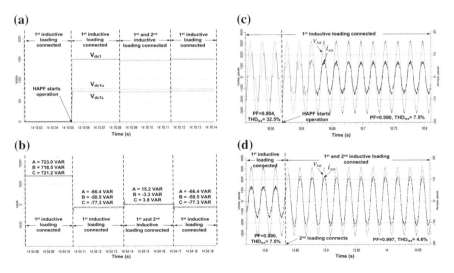

Fig. 4.72 LC-HAPF whole experimental dynamic compensation process with fixed dc-link voltage control scheme under balanced loading case: **a** V_{dc1u}, V_{dc1L}, **b** Q_{sxf}, **c** v_x and i_{sx} of phase a after LC-HAPF starts operation and **d** v_x and i_{sx} of phase a after the 2nd loading is connected

Table 4.36 Experimental results after LC-HAPF compensation with fixed dc-link voltage control under balanced loading case

After LC-HAPF compensation with fixed dc-link voltage control								
Different cases		Q_{sxf} (var)	PF	THD_{isx} (%)	THD_{vx} (%)	i_{sx} (A)	i_{sn} (A)	V_{dc1u}, V_{dc1L} Level (V)
1st inductive loading	A	−66.4	0.990	7.5	1.3	4.943	1.347	75
	B	−50.5	0.990	8.7	1.2	5.024		
	C	−77.3	0.989	9.0	1.1	5.108		
1st and 2nd inductive loading	A	15.2	0.997	4.6	1.1	8.071	1.324	75
	B	−3.3	0.997	5.2	1.0	8.036		
	C	3.8	0.997	5.8	1.1	8.023		

Figure 4.72a shows that the experimental V_{dc1U}, V_{dc1L} level can be controlled as its reference 75 V without any initial start-up pre-charging process, and be kept at its reference no matter when the 1st inductive loading or the 1st and 2nd loadings are connected. Moreover, the experimental Q_{sxf} can be approximately compensated at close to zero for both loading cases, as shown in Fig. 4.72b. Figure 4.72c shows that the experimental PF and THD_{isx} of phase a can be improved from 0.804 to 0.990 and from 32.5 to 7.5 % once the LC-HAPF starts operation respectively. From Fig. 4.72d, the experimental PF and THD_{isx} of phase a become 0.997 and 4.6 % when the 2nd loading is connected.

Table 4.36 summarizes the compensation results of the LC-HAPF based on the fixed dc-link voltage control schemes, in which both the THD_{isx} and THD_{vx} satisfy the international standards [31–33]. Compared with the experimental results before LC-HAPF compensation as shown in Table 4.32, the system current i_{sx} and neutral current i_{sn} can be significantly reduced after LC-HAPF compensation.

4.8.5.3   Experimental Results of Adaptive DC-Link Voltage Control

In a balanced loading situation as shown in Fig. 4.70 and Tables 4.32, from 4.35, the adaptive dc-link voltage level for the LC-HAPF is V_{dc1u}, V_{dc1L} = 50 V for the 1st loading and V_{dc1u}, V_{dc1L} = 75 V for the 1st and 2nd loadings. Therefore, the adaptive dc-link voltage control method for the LC-HAPF can reduce the switching loss and the switching noise compared with a fixed V_{dc1u}, V_{dc1L} = 75 V case as verified by Table 4.39 and Fig. 4.75.

Figure 4.73a shows that the experimental V_{dc1u}, V_{dc1L} can be adaptively changed according to different loading cases. From Fig. 4.73b, the experimental Q_{sxf} can be compensated close to zero for both loading cases, compared with Fig. 4.70. Figure 4.73c shows that the experimental PF and THD_{isx} of phase a can be improved from 0.804 to 0.990 and 32.5 to 8.3 % once the LC-HAPF starts operation. From Fig. 4.73d, the experimental PF and THD_{isx} of phase a become 0.997 and 4.5 % when the 2nd loading is connected. Table 4.37 summarizes the compensation results of the LC-HAPF with the adaptive dc-link voltage control scheme

Fig. 4.73 LC-HAPF whole
experimental dynamic
compensation process with
adaptive dc-link voltage
control scheme in a balanced
loading case: **a** V_{dc1u}, V_{dc1L},
b Q_{sxf}, **c** v_x and i_{sx} of phase
a after LC-HAPF starts
operation and **d** v_x and i_{sx} of
phase a after the 2nd loading
is connected

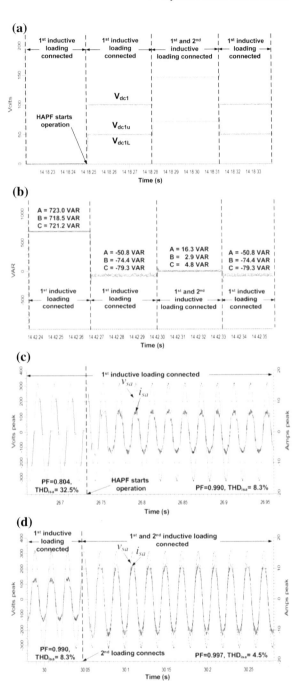

Table 4.37 Experimental results after LC-HAPF compensation with adaptive dc-link voltage control under balanced loading case

After LC-HAPF Compensation with adaptive dc-link voltage control								
Different cases		Q_{sxf} (var)	PF	THD_{isx} (%)	THD_{vx} (%)	i_{sx} (A)	i_{sn} (A)	V_{dc1u}, V_{dc1L} Level (V)
1st inductive loading	A	−50.8	0.990	8.3	1.0	5.023	1.500	50
	B	−74.4	0.989	10.3	1.1	5.044		
	C	−79.3	0.989	10.7	1.0	5.058		
1st and 2nd inductive loading	A	16.3	0.997	4.5	1.0	8.085	1.414	75
	B	2.9	0.997	5.0	1.0	8.047		
	C	4.8	0.997	5.9	1.0	8.008		

Table 4.38 Experimental results after LC-HAPF compensation with adaptive dc-link voltage control under unbalanced loading case

After LC-HAPF Compensation with adaptive dc-link voltage control								
Different cases		Q_{sxf} (var)	PF	THD_{isx} (%)	THD_{vx} (%)	i_{sx} (A)	i_{sn} (A)	V_{dc1u}, V_{dc1L} Level (V)
1st inductive loading	A	−73.6	0.990	8.8	1.0	4.891	1.319	75
	B	−55.9	0.990	9.2	1.0	5.032		
	C	−63.1	0.990	7.6	1.1	5.043		
1st and 2nd inductive loading	A	−45.9	0.995	6.4	1.1	7.100	2.629	50
	B	−56.9	0.990	9.9	1.0	5.073		
	C	−71.6	0.990	8.3	1.1	5.109		

in a balanced loading case, in which both the THD_{isx} and THD_{vx} satisfy the international standards [31–33]. Compared with the experimental results before LC-HAPF compensation as shown in Table 4.32, the i_{sx} and i_{sn} can be significantly reduced after LC-HAPF compensation.

Compared Fig. 4.73 with Fig. 4.72, the fixed and adaptive dc-link voltage control can achieve more or less the same steady-state reactive power compensation results. The adaptive control scheme just requires a lower dc-link voltage level of V_{dc1u}, $V_{dc1L} = 50$ V for compensating the 1st loading, but it will have a longer settling time during the load and dc voltage level changing situation. As discussed before, since the switching loss is directly proportional to the dc-link voltage, the adaptive dc-link voltage controlled LC-HAPF system will obtain less switching loss than the one with a fixed dc-link voltage, as verified by Table 4.39. Figure 4.75 shows the LC-HAPF experimental compensating current i_{cx} of phase a and their spectra with: (a) a fixed V_{dc1u}, $V_{dc1L} = 75$ V and (b) adaptive dc-link voltage control. Figure 4.75 clearly illustrates the adaptive dc control scheme can effectively reduce the switching noise compared with the fixed V_{dc1u}, $V_{dc1L} = 75$ V case.

Fig. 4.74 LC-HAPF whole experimental dynamic compensation process with adaptive dc-link voltage control scheme in an unbalanced loading case: **a** V_{dc1u}, V_{dc1L}, **b** Q_{sxf}, **c** v_x and i_{sx} of phase a after LC-HAPF starts operation and **d** v_x and i_{sx} of phase a after the 2nd loading is connected

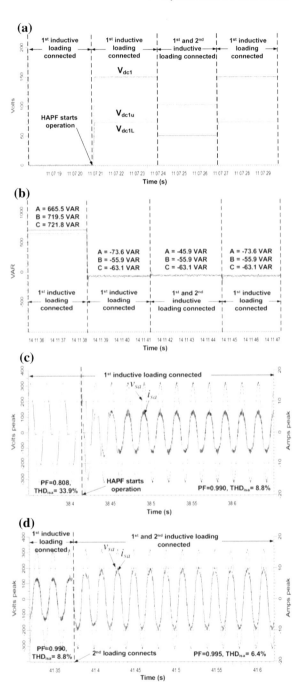

Table 4.39 Experimental inverter power loss of LC-HAPF with fixed V_{dc1u}, $V_{dc1L} = 75$ V and adaptive dc-link voltage control under balanced and unbalanced loading cases

Inverter power loss of LC-HAPF at different situations			Fixed V_{dc1u}, $V_{dc1L} = 75$ V (W)	Adaptive DC-Link Voltage Control
Balanced loading case	Power loss (W)	1st inductive loading	41	37 W (50 V) ~ 10 %↓
		1st and 2nd inductive loading	59	59 W (75 V)
Unbalanced loading case	Power loss (W)	1st inductive loading	41	41 W (75 V)
		1st and 2nd inductive loading	45	40 W (50 V) ~ 11 %↓

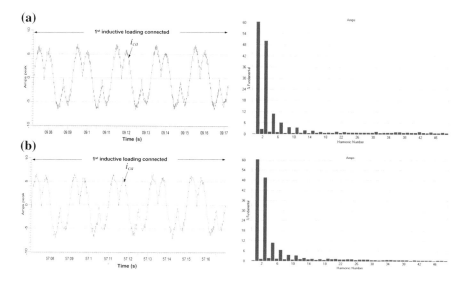

Fig. 4.75 Experimental and frequency spectrum for i_{cx} of phase a under balanced loading situation with: **a** a fixed V_{dc1u}, $V_{dc1L} = 75$ V and **b** adaptive dc-link voltage control

The adaptive dc-link voltage control can also work in an unbalanced loading case. Notice the unbalanced loading case as shown in Fig. 4.71 and Table 4.33, from Table 4.35, the required minimum V_{dc1u}, $V_{dc1L} = 75$ V for the 1st loading and V_{dc1u}, $V_{dc1L} = 50$ V for the 1st and 2nd loadings. Therefore, the adaptive control method for the LC-HAPF can reduce the switching loss (in Table 4.39) and switching noise compared with a fixed V_{dc1u}, $V_{dc1L} = 75$ V case when the 1st and 2nd loadings are connected.

Figure 4.74a shows that the experimental V_{dc1u}, V_{dc1L} can be adaptively changed according to different loading cases. From Fig. 4.74b, the experimental Q_{sxf} can be

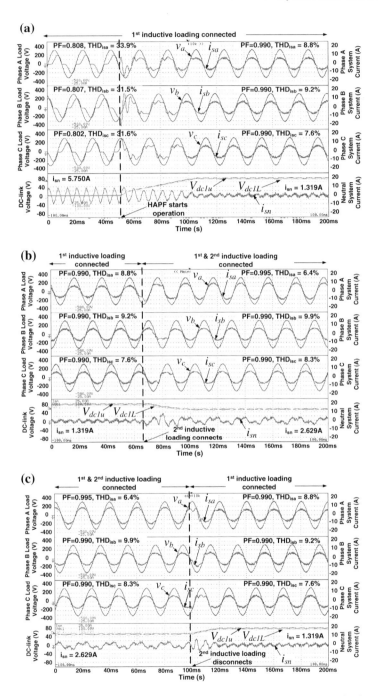

Fig. 4.76 LC-HAPF whole experimental dynamic compensation process with adaptive dc-link voltage control scheme under unbalanced loading case: **a** before and after LC-HAPF starts operation during the 1st loading connected, **b** before and after the 2nd loading is connected and **c** before and after the 2nd loading is disconnected

compensated at close to zero for both loading cases, compared with Fig. 4.71. Since phase b and c loadings are kept the same as the 1st inductive loading case under the balanced loading situation, their corresponding compensation waveforms are not included here. Therefore, only phase a compensation waveforms are given. Figure 4.74c shows that the experimental PF and THD_{isx} of phase a can be improved from 0.808 to 0.990 and 33.9 to 8.8 % once the LC-HAPF starts operation. From Fig. 4.74d, the experimental PF and THD_{isx} of phase a become 0.995 and 6.4 % when the 2nd loading is connected. Table 4.38 summarizes the compensation results of the LC-HAPF with the adaptive dc-link voltage control scheme under the unbalanced loading case, in which both THD_{isx} and THD_{vx} satisfy the international standards [31–33].

Table 4.39 summaries the experimental results for power loss. Therefore, the adaptive control method for the LC-HAPF can reduce the switching loss and switching noise compared with the traditional fixed dc-link voltage case no matter under the balanced or the unbalanced loading situations (Fig. 4.75).

4.8.5.4   Experimental Results of LC-HAPF Dynamic Compensation Capability

Figure 4.76 shows the LC-HAPF whole experimental dynamic compensation process under unbalanced loading case. The figure clearly shows the dc-link voltage can be adaptively changed according to different loading situations. The transient process is finished once the dc-link voltage reaches its adaptive reference level. The LC-HAPF can work for the initial start-up dc-link self-charging function with a response time of less than 3 cycles under the maximum dc-link voltage level of V_{dc1u}, V_{dc1L} = 75 V consideration. Figure 4.76 clearly illustrates the good dynamic compensation capability of the 220 V/10 kVA LC-HAPF experimental prototype.

In this section, a 220 V/10 kVA three-phase four-wire center-split LC-HAPF experimental prototype is designed and built. The parameters design of the PPF part, the design of transducers with signal conditioning circuits, IGBT drivers and digital control system for the active inverter part of the LC-HAPF experimental system are introduced. Different experimental results are provided to verify its effectiveness in dynamic reactive power, current harmonics and neutral current compensation. With the implementation of the adaptive dc-link voltage control, the LC-HAPF can achieve switching loss and switching noise reduction (without adding any extra soft-switching hardware components) compared with the conventional fixed dc-link voltage controlled LC-HAPF.

4.9 Summary

This chapter compares different hybrid active power filter (HAPF) topologies, and the pros and cons of each topology are presented. After that, this chapter presents and discusses the design and control study of a three-phase four-wire LC-HAPF system, and a 220 V/10 kVA experimental prototype is also designed and tested to show its validity. The key contributions and achieved results of this chapter are summarized in the following:

(1) The LC-HAPF good resonance phenomena prevention capability, filtering performance and system robustness have been investigated and verified by simulations.

(2) The influence on the dc-link voltage during LC-HAPF performs dynamic reactive power compensation is presented and analyzed. After analysis, by using the conventional dc-link voltage control methods as active or reactive current component, LC-HAPF either fails to achieve the initial start-up dc-link self-charging function or fails to provide dynamic reactive power compensation. To solve this, a dc-link voltage control method is proposed, which can achieve start-up dc-link self-charging function, dc-link voltage control and dynamic reactive power compensation. Simulation and experimental results are presented to verify all discussions and analysis, and also show the effectiveness of the proposed dc-link voltage control method.

(3) To obtain switching loss reduction function without adding extra circuit components, an adaptive dc-link voltage controlled LC-HAPF for dynamic reactive power compensation capability is addressed, in which it can achieve good dynamic reactive power compensating performance as well as reducing the switching loss and switching noise compared with the traditional fixed dc-link voltage controlled LC-HAPF, in which its validity and effectiveness are also verified by both simulation and experimental results. After that, the adaptive dc-link voltage control idea is also extended into the APF application, in which both simulation and experimental results also show significant switching loss and switching noise reduction compared with the traditional fixed dc-link voltage controlled APF.

(4) The minimum inverter capacity analysis of the LC-HAPF has been deduced via its single-phase equivalent circuit models, in which the analyses are verified by both simulations and experiments.

(5) A 220 V/10 kVA three-phase four-wire center-split LC-HAPF experimental prototype is designed and tested. The design details of the coupling LC and active inverter part of the LC-HAPF are described. The prototype was proven the ability to compensate dynamic reactive power, current harmonics and neutral current. Compared with the traditional fixed dc voltage controlled LC-HAPF, the switching loss and switching noise can also be reduced by the

proposed adaptive dc voltage controlled LC-HAPF. Moreover, it can significantly reduce the three-phase and neutral currents. In addition, the compensated current quality data satisfies the requirements of the international standards.

References

1. H. Akagi, Trends in active power line conditioners. IEEE Trans. Power Electron. **9**(3), 263–268 (1994)
2. H. Akagi, New trends in active filters for power conditioning. IEEE Trans. Ind. Appl. **32**, 1312–1322 (1996)
3. H. Fujita, H. Akagi, A practical approach to harmonic compensation in power systems—series connection of passive and active filters. IEEE Trans. Ind. Applicat. **27**, 1020–1025 (1991)
4. F.Z. Peng, H. Akagi, A. Nabae, A new approach to harmonic compensation in power systems —a combined system of shunt passive and series active filters. IEEE Trans. Ind. Applicat. **26**, 983–990 (1990)
5. L. Chen, A.V. Jouanne, "A comparison and assessment of hybrid filter topologies and control algorithms," in *Proceedings of the IEEE 32nd Annual Power Electronics Specialists Conference, PESC*. 01, vol. 2 (2001), pp. 565–570
6. F.Z. Peng, H. Akagi, A. Nabae, Compensation characteristics of the combined system of shunt passive and series active filters. IEEE Trans. Ind. Applicat. **29**, 144–152 (1993)
7. P. Salmerón, S.-P. Litrán, A control strategy for hybrid power filter to compensate four-wires three-phase systems. IEEE Trans. Power Electron. **25**, 1923–1931 (2010)
8. P. Salmeron, S.-P. Litran, Improvement of the electric power quality using series active and shunt passive filters. IEEE Trans. Power Del. **25**, 1058–1067 (2010)
9. S. Khositkasame, S. Sangwongwanich, "Design of harmonic current detector and stability analysis of a hybrid parallel active filter," in *Proceedings of the Power Conversion Conference*, vol. 1 (1997), pp. 181–186
10. Z. Chen, F. Blaabjerg, J.K. Pedersen, "Harmonic resonance damping with a hybrid compensation system in power systems with dispersed generation," in *IEEE 35th Annual Power Electronics Specialists Conference, PESC 04*, vol. 4 (2004), pp. 3070–3076
11. H.-K. Chiang, B.-R. Lin, K.-T. Yang, K.-W. Wu, "Hybrid active power filter for power quality compensation," in *International Conference on Power Electronics and Drives Systems, PEDS 2005*, vol. 2 (2005), pp. 949–954
12. V.F. Corasaniti, M.B. Barbieri, P.L. Arnera, M.I. Valla, Hybrid active filter for reactive and harmonics compensation in a distribution network. IEEE Trans. Ind. Electron. **56**(3), 670–677 (2009)
13. R. Khanna, S.T. Chacko, N. Goel, "Performance and investigation of hybrid filters for Power Quality Improvement," in *5th International Power Engineering and Optimization Conference, PEOCO*, (2011), pp. 93–97
14. S.T. Senini, P.J. Wolfs, "Systematic identification and review of hybrid active filter topologies," in *Proceedings of the IEEE 33rd Annual Power Electronics Specialists Conference, PESC*. 02, vol. 1 (2002), pp. 394–399
15. S. Senini, P.J. Wolfs, Analysis and design of a multiple-loop control system for a hybrid active filter. IEEE Trans. Ind. Electron. **49**(6), 1283–1292 (2002)
16. J.-H. Sung, S. Park, K. Nam, "New hybrid parallel active filter configuration minimising active filter size," in *IEE Proceedings of the Electric Power Applications*, vol. 147, no. 2 (March 2000) pp. 93–98

17. D. Rivas, L. Moran, J.W. Dixon, J.R. Espinoza, Improving passive filter compensation performance with active techniques. IEEE Trans. Ind. Electron. **50**, 161–170 (2003)
18. H. Fujita, T. Yamasaki, H. Akagi, A hybrid active filter for damping of harmonic resonance in industrial power systems. IEEE Trans. Power Electron. **15**, 215–222 (2000)
19. H. Akagi, S. Srianthumrong, Y. Tamai, "Comparisons in circuit configuration and filtering performance between hybrid and pure shunt active filters," in *Conference Rec. IEEE-IAS Annual Meeting*, vol. 2 (2003), pp. 1195–1202
20. S. Srianthumrong, H. Akagi, "A medium-voltage transformerless AC/DC power conversion system consisting of a diode rectifier and a shunt hybrid filter," IEEE Trans. Ind. Applicat. **39**(3), 874–882 (2003)
21. W. Tangtheerajaroonwong, T. Hatada, K. Wada, H. Akagi, Design and performance of a transformerless shunt hybrid filter integrated into a three-phase diode rectifier. IEEE Trans. Power Electron. **22**(5), 1882–1889 (2007)
22. H.-L. Jou, K.-D. Wu, J.-C. Wu, C.-H. Li, M.-S. Huang, Novel power converter topology for three phase four-wire hybrid power filter. IET Power Electron. **1**, 164–173 (2008)
23. R. Inzunza, H. Akagi, A 6.6-kV transformerless shunt hybrid active filter for installation on a power distribution system. IEEE Trans. Power Electron. **20**, 893–900 (2005)
24. V.-F. Corasaniti, M.-B. Barbieri, P.-L. Arnera, M.-I. Valla, Hybrid power filter to enhance power quality in a medium voltage distribution. IEEE Trans. Ind. Electron. **56**, 2885–2893 (2009)
25. S. Rahmani, A. Hamadi, N. Mendalek, K. Al-Haddad, A new control technique for three-phase shunt hybrid power filter. IEEE Trans. Ind. Electron. **56**(8), 2904–2915 (2009)
26. T. Demirdelen, M. Inci, K.C. Bayindir, M. Tumay, "Review of hybrid active power filter topologies and controllers," in *2013 Fourth International Conference on Power Engineering, Energy and Electrical Drives (POWERENG)*, (2013), pp. 587–592
27. H. Akagi, Y. Kanazawa, A. Nabae, "Generalized theory of the instantaneous reactive power in three-phase currents," in *International Conference on Power Electronics*, (1983), pp. 1375–1386
28. H. Akagi, S. Ogasawara, H. Kim, The theory of instantaneous power in three-phase four-wire systems: a comprehensive approach. Conf. Rec. IEEE-34th IAS Annu. Meet. **1**, 431–439 (1999)
29. V. Khadkikar, A. Chandra, B.N. Singh, Generalised single-phase p-q theory for active power filtering: simulation and DSP-based experimental investigation. IET Power Electron. **2**(1), 67–78 (2009)
30. A. Luo, C. Tang, Z.K. Shuai, W. Zhao, F. Rong, K. Zhou, A novel three-phase hybrid active power filter with a series resonance circuit tuned at the fundamental frequency. IEEE Trans. Ind. Electron. **56**(7), 2431–2440 (2009)
31. IEEE Recommended Practices and Requirements for Harmonic Control in Electrical Power Systems, 2014, IEEE Standard 519–2014
32. IEEE Recommended Practice on Monitoring Electric Power Quality, 1995, IEEE Standard 1159:1995
33. Code of Practice for Energy Efficiency of Electrical Installation, 2005, Electrical and Mechanical Services Department (EMSD), The Hong Kong SAR Government
34. S.-U. Tai, "Power quality study in Macau and virtual power analyzer," Master's thesis (University of Macau, Macau SAR, 2012)
35. S.-U. Tai, M.-C. Wong, M.-C. Dong, Y.-D. Han, "Some findings on harmonic measurement in Macao," in *Proceedings of the 7th International Conference on Power Electronics and Drive Systems, PEDS. 07*, (2007), pp. 405–410
36. A. Luo, W. Zhao, X. Deng, Z.J. Shen, J.-C. Peng, Dividing frequency control of hybrid active power filter with multi-injection branches using improved ip–iq algorithm. IEEE Trans. Power Electron. **24**, 2396–2405 (2009)
37. A. Luo, Z.K. Shuai, Z.J. Shen, W.J. Zhu, X.Y. Xu, Design considerations for maintaining dc-side voltage of hybrid active power filter with injection circuit. IEEE Trans. Power Electron. **24**, 75–84 (2009)

38. X.-X. Cui, C.-S. Lam, N.-Y. Dai, "Study on dc voltage control of hybrid active power filters," in *The 6th IEEE Conference on Industrial Electronics and Applications, ICIEA 2011*

39. C.-S. Lam, M.-C. Wong, Y.-D. Han, Voltage swell and overvoltage compensation with unidirectional power flow controlled dynamic voltage restorer. IEEE Trans. Power Del. **23**, 2513–2521 (2008)

40. B. Singh, V. Verma, An indirect current control of hybrid power filter for varying loads. IEEE Trans. Power Del. **21**, 178–184 (2005)

41. M. Aredes, J. Hafner, K. Heumann, Three-phase four-wire shunt active filter control strategies. IEEE Trans. Power Electron. **12**, 311–318 (1997)

42. M.-C. Wong, J. Tang, Y.-D. Han, Cylindrical coordinate control of three-dimensional PWM technique in three-phase four-wired trilevel inverter. IEEE Trans. Power Electron. **18**, 208–220 (2003)

43. L.H. Wu, F. Zhuo, P.B. Zhang, H.Y. Li, Z.A. Wang, Study on the influence of supply-voltage fluctuation on shunt active power filter. IEEE Trans. Power Del. **22**, 1743–1749 (2007)

44. N. Mahan, T.M. Undeland, V.P. Robbins, *Power Electronics, Converters, Applications, and Design*, 3rd Edn. (John Willey & Sons, New York, 2003

45. Featured projects technology and trend (Mitsubishi Electric., Japan, 1998)

46. Datasheets of Voltage and Current Transducer of Ke Hai Module "KT 20 A/P" and "KV 50 A/P"

47. Datasheet of DSP-TMS320F2812

Printed in the United States
By Bookmasters